Science, Society, and the Search for Life in the Universe

Bruce Jakosky

The University of Arizona Press

Tucson

Science, Society, and the Search for Life in the Universe

The University of Arizona Press
© 2006 The Arizona Board of Regents
All rights reserved
This book is printed on acid-free, archival-quality paper.
Manufactured in the United States of America
11 10 09 08 07 06   6 5 4 3 2 1

Library of Congress Cataloging-in-Publication Data
Jakosky, Bruce M.
Science, society, and the search for life in the universe / Bruce Jakosky.
p. cm.
Includes bibliographical references and index.
ISBN-13: 978-0-8165-2613-0 (pbk. : alk. paper)
ISBN-10: 0-8165-2613-3 (pbk. : alk. paper)
1. Exobiology—Social aspects. 2. Life on other planets—Social aspects.
3. Science—Philosophy. 1. Title.
QH325.J28 2006
576.8'39—dc22

2006006342

Contents

Preface

The modern view of the potential for life elsewhere in the universe has appeared only within the last decade, and the first syntheses of this perspective are just now emerging. This is exemplified by the recent appearance of a large number of books addressing the science, the creation of related university courses and textbooks, international conferences, and so on. Despite the tremendous scientific interest in astrobiology, many questions about the societal implications of this first synthesis have not yet been addressed.

My goal here is to explore these philosophical and societal issues in astrobiology, to encourage members of the scientific community to do so themselves, and to begin a multidimensional dialogue between astrobiologists, people from other disciplines such as the humanities, and the public. The issues at the heart of this discussion deal with the nature of science as an intellectual endeavor and with how the science of astrobiology is carried out, the parallel question of why we do science (why the public values it and why the federal government supports it), the philosophical and societal significance of searching for life beyond Earth and what the implications would be of finding (or of searching for and not finding) it, and the theological and religious implications of the potential for life elsewhere.

I am an active, practicing scientist in this field, so I have aimed the book initially at the group with which I am most familiar—the scientists themselves. This approach allows me to explain why these questions should be of interest to the science community and how its members can use this information in their everyday research and teaching activities. My perspective as an active scientist also allows me to meet another of my goals: to start a dialogue on the importance of these issues.

I hope that this book will interest a broader segment of society as well. My discussion of science and society, while at a relatively high level from the perspective of each discipline (such as philosophy of science or religion), uses examples from the field of astrobiology to integrate the different issues in what I hope is a readable and straightforward manner. Thus, it can be read by undergraduate and graduate students in the sciences who want to understand the broader issues of science, by students in nonscientific fields who wish to understand more about the nature of science, and by members of the public with a deep interest in the philosophical and societal issues as well as the scientific ones.

Some of my colleagues will take exception when I state that professional scientists tend to lack interest in societal issues. While I agree that many scientists are, as individuals, very engaged by these topics, the scientific community generally is not. As a rule, the issues are not discussed at scientific conferences, they do not come up in grant proposals, and they are not addressed in either undergraduate- or graduate-level courses. And while many scientists have an interest in the questions, too many are not familiar with basic ideas that are well known in the philosophy or history of science communities. This unfamiliarity sometimes appears when scientists speaking in public forums are asked why we are exploring the universe and searching for life, what the implications of the science of life elsewhere are for religion, or why the federal government should fund the research, and respond with glib answers that make them sound naïve and uninformed.

I have been working on this book for more than five years, although most of the writing took only a year or so. I spent much of the rest of the time learning about the different areas outside the sciences that relate to astrobiology and trying to synthesize what I learned with the science. I have had helpful discussions with a lot of people, and I would like to thank each of them profusely. Although they all may not remember it, I received valuable help, comments, and discussion at various stages from (alphabetically) Mike Belton, Connie Bertka, Merry Bullock, Carol Cleland, Dale Cruikshank, Kat Eason, Ned Friedman, Matt Golombek, Bill Hartmann, John Haught, Bob

Hazen, Noel Hinners, Guillermo Lemarchand, Alan Lester, Gilbert Levin, Paul Lucey, Chris McKay, Hap McSween, Jim Miller, Ellis Miner, David Morrison, Ken Nealson, Yvonne Pendleton, Roger Phillips, Joost Polak, Doug Robertson, Sean Solomon, Jeff Taylor, and Frances Westall. As much as I wish it were otherwise, they are not responsible for any misunderstandings or misstatements herein; those are my responsibility alone.

I am grateful to Sean Solomon and Wes Huntress for hosting me in 2002–2003 during my sabbatical at the Department of Terrestrial Magnetism and the Geophysical Laboratories of the Carnegie Institution of Washington, where much of the first draft was written. I also thank Melinda Conner for her excellent copyediting of the manuscript.

Finally, I would like to express my deepest thanks to Merry Bullock, who encouraged me while I was working on the manuscript and also discussed all of the issues from both narrow and broad perspectives. Her comments allowed me to sharpen and tighten the discussion in a way that makes it, I hope, a little more useful and interesting.

Science, Society, and the Search for Life in the Universe

Introduction

Astrobiology, Science, and Society

ASTROBIOLOGY IS A DISCIPLINE THAT INVOLVES TRYING to understand the origin, evolution, and distribution of life within the universe. It starts with understanding the origin and nature of life on Earth, moves to asking where else in the solar system life might be able to exist and whether life actually is present there, and continues with the potential for life (whether microbial or intelligent) to exist on planets orbiting other stars. Humans have been asking these questions for thousands of years. The different fields of study that make up astrobiology—geology, paleontology, astrophysics, planetary science, biochemistry, evolutionary biology, molecular biology, microbial ecology, and so on—have been around for, in some cases, hundreds of years. Only within the last decade, however, have the modern contributions from these disciplines come together and has our current scientific understanding of the potential for life elsewhere been developed. Recent discoveries have been made on Earth that relate to the existence and nature of life in extreme environments, to new branches of the "tree of life" unknown even one or two decades ago, and to the earliest history of life on Earth. We now recognize that there are places on the planets and satellites in our own solar system that might be able to support an origin of life or its continued existence today, or might have been able to do so in the past. And just within the past decade we have discovered planets orbiting other stars and disks of gas and dust that are thought to be in the process of accumulat-

3

ing into planets. Although the existence of these planets and protoplanetary systems had been suspected, their discovery has changed our perception of the occurrence of planets in fundamental ways.

Issues surrounding the potential for life elsewhere are at the intellectual center of the space exploration program. Recent reports on the directions of space science for the next decade, coming from both the NASA Office of Space Sciences and the National Research Council of the National Academy of Sciences (which provides independent scientific oversight of the NASA programs), have embraced questions about life in the universe as being of immense scientific interest and importance. Even a casual reading of these reports suggests that astrobiology has become the intellectual centerpiece of NASA and the space program, if not the raison d'être for its existence. The vision for space exploration put forward by President George W. Bush on January 14, 2004, emphasizes the search for life as one of the central drivers of the space program, with the idea of human exploration as a major theme.

The search for life elsewhere is one of only a few scientific disciplines that consistently garner the public's deep interest and attention. New discoveries that relate to the potential for life elsewhere are regularly reported on the front page of the *New York Times*. This interest includes all aspects of astrobiology, including a deeper understanding of life here on Earth, the nature of the planets and satellites in our solar system, and our understanding of worlds outside our solar system. A discovery of evidence that was thought to point to microbial life on Mars made front-page headlines worldwide in 1996, and even drew public comments from then-President Bill Clinton. This strong public interest is an offshoot of the deep significance that the possible existence of extraterrestrial life has to many, if not most, people. Whether this meaning relates to addressing the "Big Questions" of our existence, to the possibility of making contact with alien beings, or to the potential impact on our understanding of religion is subject to discussion. Regardless, the connection is real.

There was a time in recent memory when the exploration of space was at the center of the national zeitgeist. In the early 1960s, only forty years ago, there was public excitement, a sense of purpose, and a sense of destiny in the beginnings of the space program. Americans who are over the age of fifty can remember the countdowns of the rocket launches that sent men into space for the first time. People around the world over the age of forty can remember the first manned landing on the Moon and the first steps by humans onto another world. These events truly represented, in Neil Armstrong's words, "a giant leap for mankind." This excitement, however, was rooted in the first exploration of space by humans. The interest in the Moon was in the human drama rather than in the opportunity the landing presented to enhance understanding of another object in our solar system.

That sense of wonder and awe is generally lacking in space exploration today. The human space program no longer generates intense excitement. The search for life elsewhere, however, *is* able to generate that same level of excitement. Many Americans over the age of thirty remember Carl Sagan's *Cosmos* series on television, which managed to relate the scientific exploration of space to society here on Earth.

Certainly, the public has always been interested in knowing whether life exists elsewhere in the universe, but that alone is not enough to fuel a research program. It is the dramatic change in our scientific understanding of life that has taken place over the past decade that has brought the questions back to the forefront of space exploration and pushed them to the center of the NASA programs.

Despite their shared interest in life elsewhere, there is very little real interaction between the scientific community and the public. Some individual scientists are trying to understand, explore, and pursue the broader societal and philosophical issues related to their discipline, but scientists as a group show little interest in doing so. And they show even less interest in listening to the public's views about what questions scientists should be trying to address.

Because the tremendous public interest in astrobiology is relatively recent, there has been little opportunity to bring the deep underlying issues that stem from the science before the public. Why is it important to have a dialogue between scientists and the public? The fundamental reason is that scientists are exploring the universe in behalf of humanity. If the scientists are not addressing the questions that most interest the public, then they are not doing their jobs. Similarly, scientists who are not bringing their results back to the public, in a form that addresses the public interest and in ways that can be understood and appreciated, are not doing their jobs.

But a dialogue between the scientists and the public is not enough. Other groups have important contributions to make to this discussion as well. Astrobiology is an amalgam of existing disciplines, each with different ways of "doing science." There is no single way to define astrobiology. Scientists from one discipline often do not understand the approach taken by scientists from other disciplines, and as a result, they cannot effectively integrate results across disciplines and understand the breadth of their own field. Questions relating to how science is "done" and how we understand the causes behind observed phenomena fall within the category of philosophy of science. As such, they tend to be studied by academics in philosophy departments who may have no formal training in science itself. Most scientists pay little or no attention to philosophy of science, and thus most have little understanding of the validity of their own approaches to doing science. In a field such as astrobiology, which is comprised of such different component disciplines, there is real value in discussing the different philosophical approaches to the science.

Science and religion are often placed at opposite poles in the continuum of human knowledge. Their relationship is often seen as an antagonistic one. Many people think that the discovery of life elsewhere would have a tremendous (and probably deleterious) effect on religion. This view is not new. The religious implications of science discoveries, particularly in astronomy, go back hundreds of years, at least to the time of Galileo's arrest for the heresy of saying that the Earth goes around the Sun. What might the impact on religion be of discovering life elsewhere? What might the ramifications be of searching for life and not finding any? While the scientist clearly has much to contribute to this discussion, any understanding of the religious implications has to include analysis by theologians.

Although many scientists do not appreciate it, "exploration science" plays a large role in our society. Some planetary scientists may correctly see themselves as explorers but still fail to understand how this exploration fits into the broader scientific enterprise. Why is exploration science important in our society, especially given that there is so little opportunity for practical applications? Why does the federal government support this type of research? Astrobiologists need to understand the relationship between NASA, the federal government, and the public, and to understand as well what drives the support for the science.

There is a common thread running through the last few paragraphs: Astrobiology as science, science as a public activity, and science as public policy address important issues that are central to carrying out the science yet are seldom addressed within the science community. Scientists cannot address these issues alone; a full discussion must involve philosophers, theologians, public policy wonks, and, of course, the public itself. In a multidimensional dialogue, each group can bring its own background and perspective to the discussion so that the parts can be integrated into a whole. Issues can be addressed at the interface between science and religion, or between science and philosophy, only if each side provides its unique perspective and if the groups are willing to carry on a dialogue. If scientists are to provide their perspective, though, they need to have a perspective!

With this idea in mind, let us examine the philosophical and societal issues in science in general and in astrobiology in particular. This discussion will have two goals: first, to provide a scientist's view of the interactions between astrobiology and the broader societal issues; and, second, to convince practicing astrobiologists that ignoring these issues puts them at risk of becoming disconnected from the broader society.

These broad issues bring together two distinct cultures within the science. The first culture is the traditional scientific view: What discoveries are being made in astrobiology? How can we integrate them into a coherent understanding of the science? What are the implications for our understanding of the potential for life to exist elsewhere, either in our own solar system or beyond, and how can we carry out a research program to determine whether extraterrestrial life actually does exist?

The second culture involves the connections between science and the humanities. What is the broader societal significance of carrying out a scientific program in astrobiology? More generally, what role should science play in society and what is its actual role? In *The Two Cultures*, published more than forty years ago at the dawn of the space age, C. P. Snow described two distinct cultures that he saw around him at the time—the sciences and the humanities—and lamented that well-educated and high-level practitioners of each were woefully ignorant of the other, and that both groups and society as a whole suffered as a result. The lack of discussion of the societal issues in astrobiology today suggests that, to a great extent, this dichotomy still exists.

The relationship between these disciplines was addressed more recently by E. O. Wilson in his book *Consilience*, which argues that the sciences and the humanities can be connected or united through common concepts and principles. *Consilience* takes a scientific approach to understanding areas that traditionally fall within the humanities, such as religion, creativity in art, and consciousness, applying scientific techniques to nonscientific areas to show that this can be done productively.

Alternatively, in astrobiology, the different disciplines address different types of questions but can be brought to bear on a common problem—understanding the nature and distribution of life in the universe. This distinction is very important. Rather than insisting on a causal relationship between the sciences and the humanities, we must understand the influence each has on the other. Science should not be done outside the influence of the humanities. After all, scientists are people, they are funded by other people, and people are interested in the results; in other words, science is a truly human endeavor. Similarly, philosophical questions regarding the nature of life, how life and intelligence originated, and whether we are alone in the universe either as life or as intelligent life are based on our understanding of the sciences.

This book is intended, then, as a response to the compelling need for discussion on a number of issues that center on the science of astrobiology yet involve its connections with the broader society:

- There is a remarkable lack of connection between scientists and the public. While most scientists have accepted the now-ubiquitous calls to engage in outreach, outreach as practiced in the space sciences typically involves a one-directional talking *at* the public rather than a dialogue *with* the public; it is at best a first step. Scientists as a group do not appreciate the societal issues that underlie their work, and as a result have a difficult time valuing them or responding to them.
- The lofty goals and ideals that characterized the early era of space exploration forty years ago have disappeared and have not yet been replaced by anything of substance. NASA and the United States lack direction with regard to space. This lack of focus emerges in the continued dithering in the wake of the 2003 loss of the space shuttle *Columbia* over what the role of humans in space should be within the broader space program, and what the role of space science should be.
- There is a lack of discussion within space science of the importance of exploring the universe around us, what the role of exploration should be, and, more generally, what the role of basic research should be.

These topics will be addressed here through a discussion of the role and significance of astrobiology in the larger society. How do philosophy and religion relate to our understanding of the science of life in the universe? Astrobiology is a relatively new science, and there has as yet been very little opportunity to develop these issues. While some will reasonably see the present discussion as "philosophy lite," even a brief and preliminary discussion such as this one can serve to raise the issues. I hope that a more in-depth discussion involving academics from disciplines outside the sciences will follow.

A discussion of the societal issues in astrobiology has to start with an introduction to the scientific issues. What is it about the nature of life on Earth that suggests that there could be life elsewhere? Where in our solar system could life plausibly occur? What processes are responsible for producing the present architecture of our solar system? How did these same processes play out in other planetary systems, and thus what is the potential for life beyond our solar system? Equally important, are we deluding ourselves in believing that because life exists on Earth it must exist elsewhere, as if we really *need* to find something out there? These topics are discussed in chapter 1.

An important related question is, What is life? We have only a single example of life—that found here on Earth. How do we—and can we—extrapolate that concept to life elsewhere? What can we say about what constitutes life? How do we classify entities that are on the border between living and nonliving, such as viruses? Can we put together a "user's guide" to finding or identifying life even without having a unique definition? Chapter 2 contains a discussion of the difficulty of defining life and a rudimentary attempt to deal with these questions.

Chapter 3 addresses the question of whether astrobiology is a science. As a scientific endeavor, exobiology, a close cousin of astrobiology, has been described pejoratively as a discipline without a subject. To determine whether this statement is fair, and to understand how astrobiology operates as a science, we need to understand how science works in general and what science is. This issue touches on questions that are being asked outside the science community that relate to absolute truth, objectivity, and cultural influence in science.

Many of the subdisciplines in astrobiology do not fit neatly into the standard description of experimental science but are better described as historical science. This approach involves observing the manifestations left behind (in the geological record, for example) by past events and trying to construct narratives that describe the sequence of events that must have taken place. Many scientists, even those who practice within this field, do not understand or appreciate this distinction, yet it is fundamental to understanding how to interpret the results of their research. The nature of historical science and the value of historical narratives are explored, with examples drawn from recent astrobiological thought, in chapter 4.

The parallel question to how we do science is *why* we do science. This topic touches on key questions that include the roles of basic versus applied science, the national science policy, and personal and societal motivations. More specifically, we can address the role of astrobiology in the process of exploring the universe and the role that this exploration plays in our society. Also in this category is the philosophical significance of searching for or finding life elsewhere and what the answers imply for how we choose to construct and promote our space exploration program. These issues are addressed in chapter 5.

Would the discovery of life elsewhere have any impact on religion? A large fraction of the public thinks that it would, although most theologians are less concerned about it. Many issues dealing with the universe outside the Earth touch on religion, including the "anthropic" universe, the potential for life elsewhere, and the potential for intelligent life elsewhere. As part of this discussion it is fair to ask whether science itself can be viewed as a religion. The answer to that question, discussed in chapter 6, has important implications for how society deals with the relationship between religion and science.

Finally, in chapter 7, I revisit the two different cultures within astrobiology—the science and its societal connections. Despite the separation that exists today between basic research and exploration science and the public, it is public interest that supports space exploration. I use the idea of two cultures as a starting point to indicate ways in which the science community needs to redirect itself in order to be responsive to these issues.

To some extent, this book is aimed directly at members of the planetary science, astrophysics, and astrobiology communities; however, nonscientists should understand it as well. My goal is to begin a dialogue about what the role of exploration science is and should be in our society. I take the first step by providing a scientist's perspective and by encouraging the scientific community to address these questions, setting the tone and direction for the real discussions across disciplines that I hope will result.

1

Is There Life Elsewhere?

WE KNOW OF ONLY ONE EXAMPLE OF LIFE IN THE UNIverse: that found here on the Earth. We can predict whether there might be life elsewhere—either on other planets and satellites in our own solar system or on bodies orbiting other stars—but no matter how reasonable and plausible it might seem that life should exist elsewhere, we still won't know for sure. We can, however, examine the origin and evolution of life on Earth and apply what we learn to the relationship between terrestrial life and the planet on which it resides, the ability of planets and satellites in our own solar system to meet the requirements to be able to support life, and the likelihood that planets or satellites around other stars might meet those conditions.

We will have the potential to discover life on a planet other than Earth within perhaps the next one or two decades. We have the technological capability today to explore Mars, the rest of the solar system, and beyond. We might find life on Mars or Europa, or we might find evidence that life exists on Earth-like planets orbiting other stars. In both cases we are literally on the verge of discovering life elsewhere in the universe—if it exists.

Characteristics of Life on Earth

What characteristics of life on Earth can help us to understand whether life could exist elsewhere? The most important aspects are discussed here, starting with the components that make up life and then moving to the specific details of life on Earth.

The elemental building blocks of life are very common throughout the universe. The major elements constituting life on Earth are carbon, hydrogen, oxygen, and nitrogen (C, H, O, and N, respectively), with two dozen or so other elements also playing a role. Hydrogen was created during the earliest history of the universe, just after the Big Bang. When the temperature had cooled sufficiently, the photons of light that formed the original universe would have been converted into matter, making the elements hydrogen and helium (and trace amounts of lithium).

Carbon, oxygen, and nitrogen were created later, under extreme temperature and pressure inside stars, by the nuclear reactions that fueled the stars' production of heat and light. The stars themselves were made of hydrogen and helium gas that had accumulated together into galaxies. In the stars' centers, hydrogen atoms fused together, forming helium atoms. When the hydrogen began to run out, the star collapsed a little bit, heated up further, and began fusing three helium atoms together to form an atom of carbon. Carbon could combine with helium to make nitrogen, which could combine again with hydrogen to make oxygen.

The fusion by-products created within the stars were ejected back into interstellar space when these first stars died—either by supernova explosion or as their outer layers were ejected at the end of their lifetimes. New generations of stars formed from gas that now contained small amounts of the heavier elements in addition to hydrogen and helium. Thus, the stars formed most recently have more of these heavier elements. The heaviest elements were made in the stellar explosions themselves as they were ejected into space. All elements heavier than iron—including, for example, copper, zinc, lead, gold, platinum, and uranium—were formed by these stellar explosions. Once ejected into space, they, too, were incorporated into later generations of stars and planets.

Planets are thought to have formed from the debris left behind when gas and dust collapsed to form individual stars. The first generation of stars and planets contained only hydrogen and helium. The abundance of heavy elements increased over time, though, and the stars that formed later had enough of the heavy elements to form rocky planets. These later planets had all of the same elements out of which both Earth and life on Earth are composed.

The molecular building blocks of life are very common throughout the universe. The chemical characteristics of carbon dominate the chemistry of life on Earth. Carbon has the ability to make many different types of chemical bonds with other elements. In addition, it is very available and accessible. Here on Earth, it is present in the atmosphere as CO_2 gas, it can dissolve in substantial quantities in water (as in the Earth's oceans and lakes), and excess amounts can be stored as solid rock on the seafloor (in minerals that contain CO_2, such as limestone). Other elements, such as silicon, are also capable of forming different types of chemical bonds, but they are not as versatile as carbon. Silicon is not available in the atmosphere and dissolves in water only in minute quantities. Oxygen and nitrogen (as well as silicon) make fewer types of chemical bonds and are not as versatile structurally as carbon.

The carbon-bearing molecules that are precursors to life must have been created out of elemental C, H, O, N, sulfur (S), and phosphorus (P). We call molecules containing these elements "organic," whether they are produced by organisms or not, because of their dominance in living things. Mixing oxygen-rich molecules (such as H_2O) with oxygen-poor ones (such as methane [CH_4] or ammonia [NH_3]) in an environment in which they could react chemically would create a wide variety of molecules with intermediate amounts of oxygen. These intermediates are the organic molecules that are the building blocks for life. The formation of complex organic molecules could have occurred naturally on the early Earth in a variety of different environments—in a reducing atmosphere, for example, when sparked by lightning (as in the famous Miller-Urey experiments). They also could have been produced in the energy-rich environment of hydrothermal systems, where water circulating through volcanically heated zones beneath the surface equilibrates chemically with the subsurface rocks. When the water is mixed back to the surface or into the oceans, the dissolved chemicals are out of equilibrium with their environment, and chemical reactions can produce carbon-bearing organic molecules.

In addition, organic molecules would have arrived on the early Earth intact, contained in the debris out of which the Earth was formed. We see organic molecules today in meteorites that fall to Earth, and asteroids and comets contain them as well. Radio telescopes have identified organic molecules in interstellar gas clouds that are common in our galaxy, suggesting that processes that can combine them into organic molecules operate throughout the galaxy. Thus, we suspect that the working materials out of which life might have originated were probably abundant, both here on Earth and elsewhere in our galaxy.

Life on Earth probably arose very quickly after it became possible for life to exist continuously. Earth's geological record goes back about 4 billion years, starting about half a billion years after the planet itself formed. Information contained within these ancient rocks allows us to determine the nature of the environment at the time the rocks formed. Rocks older than 2 billion years are very scarce, however, and have been heavily altered by subsequent geological processes, so we have had to piece together a complex story from only a few rocks.

The rocks don't tell us about the origin of life itself, but in them we do find ancient fossilized single-celled organisms. Those found in rocks formed, say, 1 to 2 billion years ago are unambiguously biological in that they look identical with modern cells and contain carbon that is demonstrably of biological origin. Fossils found in rocks 2.5 to 3.0 billion years old are widely accepted as biological. Those in rocks that are 3.5 billion years old are subject to debate, although their general characteristics and the large number of likely fossils make it probable that at least some are biological. Although older rocks do not contain fossil organisms, some rocks that are 3.9 billion years old show carbon isotopic evidence that life existed at that time. In sum, although we do not know when life originated, it is likely that it already existed 3.5 billion years ago, and there is pretty good evidence that it existed nearly 3.9 billion years ago.

Life could not exist continuously on the Earth until the rain of planetesimals out of which the Earth formed 4.5 billion years ago petered out. When the largest protoplanetary objects, some as big as Mars, impacted the still-forming Earth, they would have heated the surface and atmosphere enough to evaporate any oceans, raise temperatures to 3000 K, and kill off any life that might have existed. We don't know exactly when the last "Earth-sterilizing impact" was, but we can make good statistical arguments that it probably occurred between about 4.2 and 3.9 billion years ago.

Thus, the time between the last killer impact and the time that we find the earliest evidence of life is relatively short—perhaps only 100 million years, and almost certainly less than 1 billion years, a short time compared with the lifetime of a planet or a star. This argument still holds even if life did not appear until close to 2.5 billion years ago; a 2-billion-year period for the origin of life is still short compared with the 5- to 10-billion-year lifetime of the Earth or Sun. If this short time required for life to originate is typical, the origin of life would seem to be a relatively likely event in the presence of the right environmental conditions—a consequence of the types of natural chemical reactions that can occur in a dirty planetary environment.

The environmental conditions that appear to be required for an origin of life or for its continued existence are relatively straightforward and are likely to be widespread. Based on our understanding of how life on Earth works, we think that the requisite environmental conditions include the presence of liquid water (although we can argue about whether a different liquid might work as well), access to the various elements out of which life is constructed (C, H, O, N, S, P, calcium [Ca], iron [Fe], and a number of other elements), and a source of energy that can drive metabolism. Life could have originated on any planet or satellite that satisfied these environmental criteria.

Hydrogen and oxygen are abundant in the universe, and we expect water to be widespread as well; certainly it is abundant in our solar system. It is thought to have been supplied to the inner, rocky planets via comets and larger ice-rich bodies that formed in the colder outer solar system.

The elements of which life on Earth is made are likely to be both abundant and accessible on any rocky planet that is geologically active, whether in our solar system or on planets orbiting other stars. The other rocky planets in our solar system formed out of much the same materials that formed our own Earth. The asteroids and meteorites, which are representative of the materials from which the planets formed, show this to be the case. Measurements made on the surface of Mars from spacecraft and Earth-based analyses of meteorites that we think came from Mars, for example, indicate that the red planet has all of the necessary elements to support life.

Most terrestrial organisms utilize energy from sunlight to drive photosynthesis (or eat organisms that are themselves photosynthetic). Some organisms utilize chemical energy released from the chemical reactions between water and rock. These organisms are found in hydrothermal systems, in essence turning the heat of volcanic activity into chemical energy, and in cooler environments where weathering reactions (analogous to the rusting of iron) can take place. While the energy to support life on other planets could come from their own sun, photosynthesis involves a far more complex set of chemical reactions than chemosynthesis does.

Given the probable abundance of water and chemical disequilibrium on other planets, we suspect that energy sufficient to support metabolism is likely to be available on any geologically active planet. Thus, most rocky planets are likely to meet the necessary environmental conditions to support either an origin of life or the continued existence of life.

The origin of life on Earth may have taken place in an environment that could occur on almost any geologically active planet. In addition to looking at the geological record, we can also look at the biological history of the earliest organisms as recorded in modern organisms. By examining the sequence of bases in DNA molecules of modern organisms we can determine the order in which different lineages diverged from each other over the last four billion years. Organisms with very similar sequences must have diverged relatively recently, while those with very different sequences diverged a longer time ago. By comparing the sequences in many modern organisms scientists have been able to construct a "tree of life" that shows the organisms' genetic relationships.

What is particularly notable in this tree of life is that the modern organisms that appear to have diverged least from the last common ancestor of all terrestrial life live in hydrothermal systems and utilize chemical sources of energy. This suggests that the earliest organisms may have been hyperthermophilic or chemosynthetic. It is possible, of course, that some catastrophic environmental event allowed only organisms with these characteristics to survive, and that we would thus see them as the most ancient. A late giant impact might have done this, for example, by heating the oceans to a temperature at which only the heat-loving organisms could survive.

The fact that hydrothermal systems meet all of the environmental conditions required to support an origin of life makes these locations points of special interest. The chemical reactions that take place at hydrothermal vents go forward naturally, which means that they release more energy than they use. This chemical energy could create complex molecules and then string them together into complicated structures, or it could support metabolism of primitive organisms. Equally important, this type of environment, either hydrothermal or aqueous, should be widespread on any geologically active planet that has liquid water.

The fact that life originated on Earth under conditions that are likely to occur on other planets makes life elsewhere plausible. Taken together, the characteristics described above suggest that life should form relatively easily in habitats that can support the necessary chemical reactions. Now we can begin to look outward beyond the Earth and ask whether the appropriate habitats are likely to occur on other planets and satellites in our own solar system or orbiting other stars.

Potential Habitats Elsewhere in Our Solar System

A number of planets and satellites in our solar system appear to meet the conditions necessary to support an origin and the continued existence of life. We cannot say whether these places actually have life, of course, without a systematic search.

Mars

Mars appears to meet (or to have met at various times in the past) all of the environmental requirements necessary for life. Images of the surface show features that were most likely carved by liquid water, such as branching networks of valleys on the ancient surfaces, large channels that appear to have been caused by catastrophic release of water from the subsurface, small-scale gullies on the walls of canyons and impact craters that suggest the relatively recent release of smaller quantities of water from within a few hundred meters of the present surface, and debris deposits and channels within the interior enclosed basins of impact craters that suggest that the crater interiors might have been filled with standing lakes at some point following their formation.

Chemical samples taken from martian meteorites provide direct evidence that liquid water has occurred within the martian crust. They contain minerals that could have been deposited only by water flowing through the crust, for example, and isotopic indicators that suggest liquid water as a medium by which gases could be exchanged between the atmosphere and the subsurface. Observations by the Mars Exploration Rover *Opportunity* in the Meridiani Planum region also point toward liquid water. Salts and other minerals that would have precipitated from liquid water are present in great abundance, and small-scale depositional features suggest that there were standing bodies of water on the surface at one time. In sum, evidence from various sources indicates that water flowed over the surface of Mars early in its history when the temperatures must have been higher than at present, that liquid water has appeared episodically at the surface throughout history, and that water has been present within the crust and in the subsurface throughout the planet's history.

Mars also has all of the elements necessary to support life. Oxygen and hydrogen (in the form of water) are present in the atmosphere as water vapor and in the polar caps and high-latitude regolith as ground ice, and likely have been present within the crust as described above. Carbon is present in the atmosphere as CO_2 and in the crust as carbonate minerals. Nitrogen is present in the atmosphere as N_2, and may be present in the crust as nitrate minerals. Analyses of martian meteorites that have landed on Earth and in situ spacecraft measurements show that the other elements that make up life are all present in the martian surface and subsurface materials as well.

Finally, Mars has had volcanic activity throughout its entire history, although at a lower rate than Earth's activity. Given the presence of both volcanic activity and crustal and surface water, we expect that chemical reactions between the water and volcanic rock would give off energy that could support metabolism. The release of energy can also occur at lower temperatures anywhere that water and rock come into contact with each other. Inventories of the amounts of energy available throughout Mars's history suggest that sufficient energy was probably present to have been able to support an origin of life there.

It also is possible that terrestrial life was transferred to Mars early in its history in the form of microbes contained inside meteorites ejected from Earth. We know that such an exchange of materials between planets is possible, because rocks from Mars have landed on Earth as meteorites. It is plausible that microorganisms could survive the trip inside some of these rocks and land in places where they could survive. If we do find life on Mars, it may have had its origins on the early Earth. It is even possible that life might have originated on Mars and subsequently been transferred to the Earth!

Europa and the Other Icy Satellites of Jupiter

Europa, a Moon-sized satellite orbiting Jupiter, is also likely to meet all of the environmental requirements for life. We know that its surface is made of water ice because sunlight reflected from it has absorption features at wavelengths that are diagnostic of ice. Gravity measurements indicate that this outer layer of H_2O must be roughly eighty to two hundred kilometers thick. Evidence suggests that this surface layer may be frozen as ice down to a depth of only five to ten kilometers and that the deeper layers could be liquid water.

Europa's surface is very lightly cratered, indicating that geological processes have been resurfacing the satellite within the last few tens of millions of years. In addition, there are features that give the appearance of movement of blocks of ice, reminiscent of the breakup of an ice sheet atop a liquid ocean. Arcuate cracks in the surface suggest the daily movement of a thin layer of ice on top of liquid water corresponding to tides. These features are best explained if the ice layer is relatively thin and underlain by a deep liquid ocean.

Europa's magnetic field appears to be "induced" by the changes in the magnetic field Europa experiences as it orbits Jupiter. The nature and strength of this induced magnetic field require the existence of an electrically conducting material within Europa's outer layers. Liquid water that has salts dissolved in it (providing conducting ions) is the only plausible material for this conducting layer. The possible presence of salts on the surface inferred from reflectance measurements is consistent with this view.

In addition to liquid water, we expect that sufficient energy would be available on Europa to support life. Europa's density and gravity field indicate the presence of rocky material beneath the ocean. The energy could be available at the boundary where the rocky interior is in contact with the water (and, again, involving chemical reactions that can take place between the water and the rock). Energy also could be supplied from the surface. Complex molecules produced by the interaction of Jupiter's radiation environment with Europa's icy surface could be incorporated into the ice and eventually transported into the ocean; once there, chemical reactions involving these molecules would give off energy. Europa does not seem to have as much chemical energy as is available on Earth or Mars, but we imagine that it might have been able to support an origin of life.

The Jupiter satellites Ganymede and Callisto are also thought to contain a substantial amount of water ice—about half of the volume of each satellite. Their surfaces do not show evidence of active geological processes as recent as those on Europa that led us to believe that there could be liquid water beneath the icy surface. Both show the same type of induced magnetic field that we saw on Europa, however, that can be explained only by the presence of a global-scale liquid water ocean with dissolved salts in it. The same arguments made for Europa regarding the availability of the elements to support life and energy to support metabolism apply here as well. If there is life on Ganymede or Callisto, though, it is buried deep below the surface, and we may never be able to access it or demonstrate its existence.

Titan

With an atmosphere thicker than Earth's, and clouds and haze that obscure the surface from view, Saturn's satellite Titan once seemed to have the potential for a warm climate and an extensive biosphere. Astronomers imagined that the dense atmosphere could have provided a greenhouse warming that would have raised surface temperatures high enough to melt water ice. The atmospheric haze is rich in organic molecules formed from methane and ammonia, and complex organic molecules may have rained out of the atmosphere and covered the surface with a kilometers-thick layer. Today, we know that Titan is too cold to allow liquid water to exist; there is little greenhouse warming. Even though liquid methane and ethane lakes, oceans, or "groundwater" may be present, temperatures are too low for significant chemical reactions to have proceeded very far toward life.

On the other hand, Titan probably is covered in part by water ice that would have been melted by the heat from occasional impacts of asteroids and comets onto the surface. Small lakes would have existed briefly and been exposed to the atmosphere. It would be interesting to see what types of chemical reactions might have occurred there, and how far along the path toward life these chemical reactions might have proceeded.

Venus

Just as we wish to understand where life might exist and what makes some planets and satellites habitable, we also wish to understand why others are not habitable. If we understand the different outcomes that can occur in planetary evolution, we can extrapolate our knowledge of our own solar system to planets orbiting other stars.

Venus, the second planet from the Sun, does not appear to be habitable. It is a rocky planet about the same size and mass as the Earth, with a similar inventory of CO_2 and N_2, and it probably formed from a similar complement of elements. But Venus has a thick CO_2 atmosphere and a surface temperature of about 750 K. Radar imaging of the surface shows some similarities with Earth (active tectonic activity, abundant volcanism, aeolian processes) and some significant differences (no global-scale plate tectonics, greenhouse warming of hundreds of degrees, no liquid water at the surface).

Venus had more water in the past than it has today. The water was lost when water vapor in the atmosphere was broken apart by sunlight into hydrogen and oxygen atoms and the hydrogen atoms escaped into space. Because hydrogen would escape more efficiently than deuterium (D, a heavier form of hydrogen with an extra neutron in its nucleus), the atmosphere became enriched in the latter. This enrichment in the ratio of D to H remaining in the atmosphere (enriched relative to Earth's water by a factor of 135) demonstrates that water was indeed lost to space.

The atmosphere of Venus is thought to have reached its current state as a result of a runaway greenhouse. Something similar would happen if the Earth were moved into Venus's orbit. Closer to the Sun, it would heat up a little bit. The increase in temperature would cause some water from the ocean to evaporate into the atmosphere. Water vapor is a greenhouse gas, so its addition to the atmosphere would increase the greenhouse trapping of heat and raise the surface temperature a bit more. This would cause a little bit more water to evaporate, and so on, in a positive-feedback loop that would not end until all of the water from an ocean was in the atmosphere and temperatures were extremely hot. The heat would have broken apart any carbonate-bearing minerals at the surface and released all of their CO_2 into the atmosphere as well. Even after the water vapor was broken apart and the hydrogen lost to space, the CO_2, which is also an effective greenhouse gas, would remain in the atmosphere and temperatures would stay high.

Four billion years ago, however, the output from the Sun probably was about 30 percent lower than it is today, and Venus might have been habitable then. At that time, an Earth-like planet Venus's distance from the Sun might not yet have become a runaway greenhouse. Only as the Sun heated up would Venus have made the sudden transition from relatively clement conditions to a runaway greenhouse atmosphere.

Our inner solar system may thus have had three habitable planets in its earliest history, four billion years ago—Earth, Venus, and Mars. Life might have originated on any one of them, and all three might have exchanged organisms via the impact ejection of rocks!

What's the Bottom Line?

We have no proof that any planet or satellite in our solar system other than Earth has life today or ever had life. The arguments in favor of life are based on our views of the origin and evolution of life on Earth, the environmental conditions required for life, and our understanding of the actual conditions that occur there. Together, they tell us that it is possible, and indeed plausible, that life exists or once existed elsewhere in our solar system. And they constitute a justification for exploring these worlds with the goal of determining whether life is present.

The answer—whether there is life on Mars or Europa, for example—will tell us much about the possible abundance and distribution of life in the universe. Failure to find evidence of life, even though these worlds seem to meet the environmental conditions necessary for it, would suggest that our understanding of the origin and evolution of life on Earth is incomplete. It would mean that an origin of life may not be as easy as we currently think, and that life might not be as widespread throughout the galaxy and universe as we might expect.

If we do find life on Mars and Europa, though, and if it represents an origin independent of life on Earth, that would indicate that the origin of life must be fairly straightforward. If habitable planets and satellites are abundant in the galaxy, as we think they are, then life itself would likely be very widespread as well. From the scientific perspective, either answer—yes or no, life or no life—will have major ramifications for our understanding of the distribution of life in the universe.

The Potential for Life Outside Our Solar System

Is it possible that complex and possibly intelligent life exists outside our solar system? Recent discoveries offer some insight into the scientific issues regarding the existence of life beyond our own solar system.

Planets have been discovered orbiting nearby stars. While our theories of planet formation suggest that planets should exist elsewhere, an ounce of evidence is much more convincing than a ton of hypotheses. The first real extrasolar planet was detected in 1995, and new ones are continually being discovered. We now know of more than 150 planets orbiting other stars. The planets generally cannot be imaged or observed directly; they are discovered primarily by the gravitational wobble that they induce on their central star. As a result, the planets that have been discovered are all massive gas giants similar to Jupiter and Saturn in our solar system. We have no direct evidence for the existence of Earth-sized or Earth-like planets (with one exception that is not of interest for possible life).

Almost all of the planets that have been detected are oddballs compared with those in our own solar system. Most either orbit very close to their star (many of them closer than Mercury is to the Sun) or occupy very elliptical orbits. We don't know if this type of solar system is more typical or more representative of solar systems in general than is ours or whether this type of solar system is simply much easier to observe. Only a couple of the 150 or so planetary systems have their "Jupiters" located in the outer part of the system.

Roughly 5 to 10 percent of stars are thought to have giant planets orbiting them. The remaining 90 to 95 percent of stars may have planets as well—gas giants that have not yet been detected or Earth-like rocky planets that are too small to detect today—but we really have no information about them.

Earth-like planets probably exist, although we cannot predict their abundance. We think that the gas giant planets formed in the outer portions of the gas and dust debris disks that were left behind as their central stars accumulated. Debris in the inner portions of these disks should have consolidated into Earth-like rocky planets closer in to their stars; however, the oddball orbits of most of the known gas giants may make that idea problematic. The gas giants are thought to be able to form only in the outer solar system, where the condensation of water ice allows relatively massive ice-and-rock cores around which gas can accumulate. Closer to the star, temperatures are too warm to allow ice to condense, and any rocky core that could form would be too small to accumulate and hold on to much gas. The giant planets orbiting near their suns may have migrated inward toward their central star as a result of drag from gas remaining in the debris disk, from the accumulated dynamical effects of gravitational scattering of smaller objects, or perhaps from gravitational close encounters with other gas giants. As the planets migrated inward, they most likely would have either gobbled up or gravitationally ejected any Earth-like planets they encountered. Earths probably could not survive within these systems. It seems likely, then, that most of the planetary systems that have been discovered do *not* have Earth-like planets that could harbor life.

Habitable *satellites* could orbit some of these extrasolar planets, though, much as Europa orbits Jupiter, and they might contain a global ocean beneath a covering of surface ice. As yet we have no way of knowing whether satellites are common in other planetary systems and, if they are, whether they could survive the inward spiral through the inner solar system.

Habitable worlds may be abundant in the galaxy, and life may be widespread. We expect that any Earth-like planets would have formed in the inner portions of their stars' planetary systems, much as the terrestrial planets formed in the inner part of our solar system. Gas and debris disks such as the one that was left over when the Sun formed have been observed in the vicinity of about one-fourth to one-half of newly formed stars, and may be ubiquitous. Some of these disks even have gaps suggesting that planets might be forming within them, although we cannot see the planets directly.

If rocky planets do form within the inner regions of planetary systems, then many of them are likely to be in the "habitable zone"—the region close enough to the star to have a surface sufficiently warm for ice to melt, yet far enough away to prevent all the water in the atmosphere from evaporating in a runaway greenhouse. The habitable zone in our own solar system extends from just inside the orbit of Earth to near the orbit of Mars, a relatively wide region. In a planetary system with even just a few rocky planets, at least one would be likely to reside within the habitable zone.

Even the planets that lie outside the habitable zone in their solar system could be habitable kilometers below the surface, where temperatures are warmer and liquid water could exist. On Mars today, for example, temperatures warm enough for ice to melt can be found only a kilometer or two beneath the surface. Organisms live in just this type of subsurface environment on Earth, and there is no reason to think that they could not live in similar environments on other planets.

Finally, the gas giants orbiting other stars might have regular systems of satellites orbiting them just as Jupiter and Saturn do. These satellites might be habitable in much the same way that Europa, Ganymede, Callisto, and possibly Titan might be habitable.

If life can originate as easily as we think it can, then these habitable worlds elsewhere may have life on them. If most stars have planets orbiting them, and if even one planet or moon in each planetary system is habitable, life could exist on a hundred billion worlds in our galaxy.

Extraterrestrial intelligence could be rare or widespread. The preceding discussion of habitable planets refers generally to the potential to support microbial life—single-celled organisms that might be analogous to the simplest forms of life on Earth today. Even if microbial life is widespread throughout the galaxy, though, we don't know how likely it is that intelligence could arise. Although it took a relatively short time for life to originate on Earth, it took another couple of billion years for multicellular organisms to arise, and then nearly another billion years more for what we think of as intelligence (that is to say, humans) to come about—about four billion years in all. The long times required for these events suggest that they are more difficult transitions to make and therefore less likely.

In other words, the time required for the development of intelligence on Earth is more comparable to the lifetime of a star than to that of a habitable planet. If this time is at all representative of that required for the evolution of complex beings and intelligence, intelligence may not be nearly as widespread as microbial life; only a fraction of planets might survive long enough for intelligence to evolve.

The likelihood that intelligence will arise even on long-lived planets is impossible to determine. The factors that governed its origin here on Earth are not understood, so it is difficult to extrapolate to other worlds. There may be a general imperative toward the development of intelligence, or it may be a random event that occurred once on Earth but is unlikely to occur elsewhere.

Are We Deluding Ourselves?

The above discussion supports the hypothesis that a simple form of microbial life could develop on any geologically active planet. If there are indeed a large number of Earth-sized and potentially Earth-like planets orbiting other stars, as we believe there are, then life could be widespread throughout our galaxy.

It is somewhat worrisome, however, that we have reached this grand conclusion by extrapolating from the single example of life that we know: life on Earth. Are we deluding ourselves into justifying a conclusion that we profoundly want to be true? Is this apparent desire to find life elsewhere clouding our ability to reach appropriate conclusions from limited data? In concluding that life could be widespread, have we extrapolated far beyond what the data suggest?

Asking questions about the origin and evolution of life is not as simple as, say, predicting an orbit from known physical principles. Myriad unknown and unpredictable events can affect the history of life on a planet. It is essentially impossible to generalize from our sample of one to the nature of life in general, to understand the complex and possibly multivariate drivers behind the formation of life, or to predict the ways in which evolutionary pressure exerted on complex ecologies can result in different outcomes. We can understand how some of these properties work on Earth by studying the evolution and development of species or ecosystems. But the tremendous complexity of the systems as well as our lack of knowledge of the underlying circumstances prevents the accurate prediction of outcomes. We need to know a great deal more before we can make such predictions: Do the principles and properties that govern life on Earth operate in the same way everywhere in the universe? Does all life have to be based on DNA and RNA? Does all life have to use the same two dozen or so amino acids, out of the much larger number that might have been used? Does all life have to be based on the chemistry of the carbon atom?

The fact that life appeared so quickly here makes it hard to believe that an origin of life is an extremely difficult process. As the sportswriter and short-story author Damon Runyon said, "The race may not always go to the swift, nor victory to the strong, but that's the way you bet!" The origin of multicellular organisms and then intelligence took so long by comparison that it is equally hard to believe that these are easy or straightforward processes.

This logic does not prove that life exists elsewhere or that all origins of life are rapid, but it does provide an intellectual framework for an understanding of whether conditions that can support life exist elsewhere, whether habitable planets are widespread or rare (or somewhere in between) in our galaxy, and how we might search for life either in our own solar system or beyond. It provides a foundation that can govern our exploration.

Exploration to Understand Habitability and Life

Although we can speculate endlessly about whether life might exist on Mars or on planets orbiting other stars, the answers will not come from armchair exploration. Only actual exploration will determine whether there is microbial life or intelligent life elsewhere. Such exploration could take the form of sending spacecraft to the other planets in our own solar system to make observations and collect samples that could be returned to Earth for further analysis. Or it could involve making telescopic observations of other systems to detect and characterize Earth-like planets and look for evidence that life might exist there, or continuing our search for signals sent by intelligent beings elsewhere in our galaxy or in the universe. We can answer the question that humans have asked since they first gazed at the stars, but only by observation and experimentation. Significantly, we are on a path today that allows us to pursue this exploration. NASA is actively planning spacecraft missions — some potentially in cooperation with the space agencies of other countries — to explore the potential and actual distribution of life both in our solar system and beyond.

Spacecraft are being sent to Mars to gather data that will allow us to understand the planet's environment, its history, and its ability to support or to sustain life. Plans are being made for missions that will explore the martian environment at specific locations on the surface and either search for evidence that life might once have existed or bring samples back to Earth for further study. The detailed mission plans are being updated continually to take into account both costs and the new results that are being obtained from ongoing missions.

Europa, with its high probability of having an ocean, is high on scientists' lists to explore as well. We want to try to understand the environments that exist there and their history, and then to determine the extent to which chemical processes might have led to life. Plans call for a Europa orbiter that will first determine whether a global-scale ocean exists today and then make detailed observations on and possibly beneath the surface.

Other planets and satellites in our solar system are either less likely to harbor life or too difficult to sample. We are nevertheless exploring them to understand why they are less habitable. Only by understanding how the present-day state of these planets arose and what processes were responsible for the architecture of our solar system can we extrapolate to other planetary systems.

We are designing Earth-orbiting spacecraft that will be able to look for planets orbiting stars outside our solar system and, eventually, characterize them. If we can detect Earth-like planets and measure the properties of their atmospheres and surfaces, we might be able to detect chemicals suggesting the presence of life.

These active and planned spacecraft missions constitute a vigorous and exciting program of exploring the astrobiology of the universe. Through them we will gain a much better understanding of what planetary environments are conducive to the presence of life and how planetary formation processes lead to habitable planets—in essence, what makes habitable planets and what makes a planet habitable. We are in the midst of a revolution in our understanding of the nature of life in the universe, and the next two decades will give us new information that will certainly change our views of the worlds around us.

2

What Is Life?

SHOULD WE HAVE A SOLID DEFINITION OF LIFE BEFORE WE begin discussing the potential for life to exist off the Earth? Without one, how can we know what we are looking for or if we have found it? As we shall see, however, it isn't easy to come up with a definition that includes everything that we think of as being life, excludes everything that we think of as not being life, and also provides guidance regarding what to look for on another planet. While it is tempting to put off the discussion of the nature of life to the very end—or not to discuss it at all—trying to address the question will allow us to put many of the issues that we consider in chapter 1 into the proper context. It will also highlight the problem of determining which characteristics of life on Earth might be generic to life everywhere and which might be specific to terrestrial life.

Characteristics of Life

We can start by describing some of the characteristics that we usually think of as setting living things apart from nonliving things. We generally agree that life has order or structure that sets it apart from its surroundings; takes in nutrients and gives off waste products; utilizes energy; grows and develops; carries out specific biochemical reactions; responds to its environment; reproduces; and can adapt to its surroundings by what we describe as Darwinian evolution. Can these qualities together provide a useful definition of life?

Life Has Order or Structure

Living organisms on Earth are not made up of a random assortment of elements assembled in a random fashion. All life contains certain key elements. Carbon, hydrogen, oxygen, and nitrogen are the most central, but another two dozen or so play important roles as well. These other elements include sulfur, phosphorus, calcium, iron, and magnesium but not uranium, beryllium, or titanium. Thus, at its most basic level, life has structure by having specific elemental components that are present in roughly consistent and uniform relative abundances.

At a somewhat larger scale, these elements are organized into specific molecules. Admittedly, a very large number of these molecules are important, but an even larger number are not present in terrestrial life. Among the key molecules involved in life are ATP (adenosine triphosphate), nucleic acids, amino acids, and proteins (which are assembled from amino acids).

At an even larger scale, these molecules are themselves assembled into structures that are necessary components of terrestrial life. These structures include the membranes that separate the inside of a cell from the outside; DNA and RNA molecules, which contain genetic information and govern reproduction; and cellular structures such as chloroplasts (the photosynthetic organelle in plants) and mitochondria that carry out specific chemical functions within a cell.

A well-defined order or structure alone cannot define life, however. Minerals and rocks have order and structure in their elemental composition, in the form of specific molecules that are made from the same elements found in living things, and in the composition and structure of individual crystal grains. Similarly, large-scale geological features such as volcanoes, earthquake faults, and other geological constructs or edifices have order and structure, and we recognize these entities as not being alive.

Life Takes in Nutrients and Gives Off Waste Products

We are used to thinking of animals as taking in oxygen, using it in chemical reactions with ingested organic molecules (food), and giving off the carbon dioxide that forms during these reactions. Plants take in carbon dioxide as well as other elements and molecules from their surroundings and give off oxygen. Certain kinds of bacteria take in iron atoms from the rock that surrounds them and give off a more oxidized form of iron, and others do the same thing with manganese.

By itself, however, this characteristic is not a good discriminator of life, because many nonliving entities do the same thing. Rocks "take in nutrients" by absorbing oxygen from the atmosphere in chemical reactions that oxidize the rocks' minerals; these oxidized minerals can be "given off" as waste products in the form of mineral "weathering products." In fact, the net chemical reactions in this weathering are the same ones that provide energy to some organisms, so even the specific chemical reactions cannot be used to distinguish living entities from nonliving ones. Earth's atmosphere makes this sort of exchange as well, taking in sunlight that powers photochemical reactions in the atmosphere and giving off chemical by-products that can precipitate into solid particles that are removed from the atmosphere by rain. Fire takes in nutrients (oxygen and fuel) and gives off waste products (soot, smoke, carbon monoxide, carbon dioxide, and water).

Life Utilizes Energy

Living organisms use the energy derived from chemical reactions between nutrients to power other chemical reactions. The energy can be stored in the molecule ATP and then released when it is needed. It can be used to carry out specific functions, such as building molecules during growth (through chemical reactions that produce organic molecules as the products), or it can be utilized to carry out mechanical work (moving muscles, for instance).

Living organisms utilize energy, but nonliving things do so as well. Beaches, for example, utilize energy to "grow" in the sense that wave energy redistributes sand grains, changing the shape of the beach, and also makes additional grains of sand out of larger rocks, and thereby creates more beach.

In contrast, some things that we think of as living may not utilize energy at a given moment. Dormant spores or seeds neither take in nutrients nor give off waste products. Admittedly, they will take advantage of the right environmental conditions to do these things, but observing a spore or seed under the wrong conditions might lead one to conclude that it was inanimate and not alive.

Living Organisms Grow and Develop

Plants and animals utilize energy to create additional physical structure and get bigger. Microbes and individual cells within a multicellular organism can get bigger as well, but they also utilize energy to create the molecules that allow them to divide into two microbes or cells. Is this a defining characteristic of life?

Inanimate objects such as mountains also grow, driven by the geological process of plate tectonics; rubble piles at the base of mountains grow as weathering breaks the mountains apart; sand dunes grow if there is a supply of sand; and the federal debt grows no matter what we do. Fire grows by taking in nutrients that are used in chemical reactions, creating heat that causes the fire to expand. As is the case with many living things, growth of nonliving things can continue as long as there is fuel to provide a source of energy.

Life Carries Out Specific Biochemical Reactions

All living organisms on Earth carry out many of the same biochemical reactions. Every organism uses the same two dozen amino acids to construct proteins, uses the molecules ATP and ADP to store and release energy, and uses RNA and DNA molecules to store genetic information and to code for the construction of proteins. This list of reactions comes as close as may be possible to defining terrestrial life. If these molecules are present and their associated chemical reactions occur, then the entities in which they occur are alive; if they are not present, then the entity is not alive. It is not clear that these molecules are required by life in general, however. We don't know if these specific biochemicals are the only possible solutions to the problem of storing energy and information. Would life elsewhere absolutely have to consist of the same molecules, or is it possible that other molecules, either slightly different in structure or completely different, could carry out these functions?

Molecules with different structures that are theoretically able to carry out the same functions have recently been constructed in laboratories. This suggests that life elsewhere could have stumbled onto alternative molecules, and that RNA, DNA, and ATP might not be universal indicators of life. In addition, scientists have hypothesized a form of life based on only the RNA molecule, which could have carried out both the DNA functions related to reproduction and the protein functions of catalyzing chemical reactions. This biological system is termed the "RNA world," and it is thought to have predated the present "DNA world" here on Earth. Therefore, it seems likely that "life as we know it" does not have to be the only conceivable type of life and that other molecules could carry out the necessary functions.

Life Responds to Its Environment

When we get cold, we put on a coat; if we don't have a coat, we shiver. Plants can turn toward sunlight to maximize their intake of energy for photosynthesis. Microbes can detect gradients in the composition of the medium in which they reside and can move to take advantage of improved habitability (e.g., based on temperature, salinity, etc.) and a greater supply of food.

But nonliving things also respond to their environment in different ways. Most solids expand in size or volume when heated and contract when cooled. In some cases, rocks break ("reproduce") when heated rapidly. Rocky soils overturn in response to repeated freeze-thaw cycles, "heave" rocks to the surface, and move them around to form polygonal patterns. Beaches and sand dunes form different types of patterns in response to different intensities of waves or wind.

Life Reproduces

Living things reproduce. Some do so sexually, combining the characteristics from two different organisms. Plants produce multiple seeds, each of which can grow into a complete organism, and rhizomes, which grow underground and sprout upward into new individuals and can grow independently from broken parts. Single-celled organisms reproduce asexually, typically by duplicating internal organelles, molecules, and structures and then splitting in half.

But some living things do not reproduce; the mule—the sterile hybrid of a horse-donkey cross—is a classic example. Some individual organisms are incapable of reproducing because of genetic or physical defects. We don't imagine that they are nonliving entities because of this failure (imagine trying to tell your infertile cousins that they are not alive), so we create a way around this problem—we suggest that the individual cells that constitute a mule or some other infertile being do reproduce and are alive, so therefore the entire organism must be alive.

Some nonliving things do reproduce. Sand ripples reproduce as a result of the physics of sand movement by the wind, for example, with one ripple spawning additional ripples downwind.

Many of the counterexamples that I have mentioned come from the world of geology, for the simple reason that rocks are the major category left on Earth in the absence of biology. However, one can imagine additional problematic entities. If a robot contained the instructions to build an identical robot, would this ability to reproduce make it alive?

Life Adapts to Its Environment

Evolution by modification and natural selection—what we now call Darwinian evolution—is central to life on Earth. The concept of Darwinian evolution stems from a couple of simple concepts. First, there is always natural variation of characteristics within a population. These are driven in part by mutations that are due to errors in copying DNA and RNA molecules or to changes induced by cosmic ray collisions that induce changes in atomic bonds that affect the functioning of the molecule that was changed. Such changes will leave some organisms better able to survive within their environment and others (most, actually) less able to survive. If these changes are incorporated into the genetic material of the organism, they are passed on to offspring.

Second, organisms typically produce more offspring than the environment can support, so that many offspring do not survive to adulthood. Thus, there will be competition among the individuals in a given population of organisms. Those that are better able to survive will produce more offspring, and by virtue of retaining the genetic information from their parents these young also will be better able to survive. Those that are less able to compete and survive will give rise to fewer offspring; their own functionality may be impaired so much by the mutation that they die without having been able to reproduce at all.

Over time, the positive characteristics will spread to the genetic material of the entire population. The accumulation of a sufficient number of positive changes eventually produces a population that is so different that it is considered a new species. Over long periods, as multiple selective events split off groups of organisms from each other, this process created the multitude of species that populate the Earth today.

This concept of change and natural selection, first described by Charles Darwin in *On the Origin of Species* (1859), is the single most important unifying theme in biology. As one biologist put it, nothing in biology makes any sense except in the context of Darwinian evolution. It is so fundamental that one possible definition of life centers on it. In a NASA report, Gerald Joyce, a molecular biologist working on biochemical reactions that can take place in the laboratory, defined life as "a self-contained chemical system capable of undergoing Darwinian evolution." Darwinian evolution is the basis of this definition; the other points are secondary. The idea of "self-contained" precludes, for example, a laboratory environment in which a key step is supplied by outside agents (e.g., humans). And a "chemical system" precludes life based on a computer program that resides within a computer or our earlier hypothetical example of robots building robots.

The biggest problem with Joyce's definition is how to determine whether something is capable of Darwinian evolution. The definition applies to populations of organisms over long periods; it cannot apply to a single organism, nor can it apply to a population observed at a single instant. Our example of a mule rears its ugly head again here, so to speak. It satisfies our intuitive definition of being alive, but because it cannot reproduce it is not capable, even in theory, of Darwinian evolution. Or is the problem moot because a single infertile individual is still a part of the population, and it is the population that is capable of Darwinian evolution?

In addition, there is the problem of "non-Darwinian evolution." Organisms are capable of exchanging DNA with other organisms. Such "lateral gene transfer" allows for sudden relatively large-scale changes in the genetic information of organisms and in their functional capability. Does this type of change lie outside Darwinian evolution, and does it affect our definition of life?

Early in the history of life, organisms may not have been independent entities, each encased in its own membrane. An organic soup of biomolecules in liquid water, each reacting both with other molecules and with its surroundings, might have been the first life. In such a case, evolution might have operated at the molecular level rather than at the organism level. Does this process lie within our definition of life?

Problems with Definitions

Unfortunately, every characteristic that we have raised for possible inclusion in a definition of life has been found wanting. For each there is a counterexample that, while admittedly somewhat contrived, shows the limitations of the definition. If we use these characteristics to construct a definition anyway, then we have tuned the resulting definition solely to be consistent with our preexisting biases toward what constitutes life on Earth. In other words, we will have carefully defined life so that it includes all of the things here on Earth that we think should be included and excludes all of the things on Earth that we think should be excluded. In this sense, our definition is fundamentally no different from a long list that includes on it every terrestrial organism that we think is alive.

Perhaps this is the wrong approach altogether. Our goal in defining life is not to be able to categorize living things here on Earth; we don't need a definition for that. Rather, we wish to examine samples from another planet and determine whether they contain evidence for life within them. By tuning our definition to terrestrial life, however, we are a priori limiting ourselves to identifying or finding life that is very similar to terrestrial life. Life that might have a different biochemistry, might multiply or reproduce on very different time scales, or might in a snapshot analysis be indistinguishable from natural, nonliving entities could not be identified in this way.

Carol Cleland, a philosopher of science at the University of Colorado with an interest in the attempts to define life, recently pointed out that, with a single example of life—that on Earth—we have the same problem scientists faced in trying to define water prior to the existence of molecular theory. We know today that water is uniquely and absolutely defined as the molecule described by the formula H_2O. Prior to understanding the nature of molecules, however, water was defined based on what it could or could not do— how it would interact with other chemicals, what it looked like or tasted like, and so on. In essence we are trying to do the same thing in using the characteristics of life on Earth to arrive at a unique definition of life.

Without having a sample of life that originated independently of that on

Earth and that might show a wider range of characteristics, molecular structures, or biochemical compositions and reactions, it is not possible to determine which characteristics of terrestrial life are specific to Earth life and which are general to all life. Thus, we cannot come up with a singular and unique definition of life.

The Origin of Life: The Same Problem Revisited

We have the same problem in trying to identify a unique boundary between living and nonliving entities on Earth some 4 billion years ago. The geologic record is too sparse to allow us to determine the processes that were associated with the origin of life. Let us imagine, however, that we could have watched the origin taking place over many hundreds of millions of years. Early on, we would have seen a system that included the Earth's surface, oceans, and atmosphere. Undoubtedly, significant chemical reactions were taking place, given the wide variety of molecules that would have been supplied from space during Earth's formation, from the interior of the Earth as a result of the differentiation and outgassing that took place during its earliest history, and from chemical reactions that were taking place in diverse chemical environments. Places of interest might have been the oceans themselves, Darwin's "warm little pond" where chemicals eventually might have come together to create life, or hydrothermal systems where heat from volcanoes or from large asteroid impacts would have driven circulation of water and created environments where disequilibrium chemical processes might have been important. Life would have been completely absent, and this environment clearly would have been dominated by chemistry and geochemistry. (If my description of the ongoing chemical reactions sounds too biological and you doubt that there was no life at this time, simply back up another 100 million years; continue to back up until you are convinced that the system was nonbiological.)

If we jump ahead a half-billion years or so, to perhaps 3.5 billion years ago, we find convincing evidence in the rock record that life existed at that time. We see structures in certain kinds of rocks that are interpreted as fossilized cells, and we see larger-scale structures that appear to have been made in environments that consisted of colonies of fossilized cells. In addition, there is isotopic evidence suggesting that life was present. (If you doubt that life existed at this time, simply jump forward in increments of a half-billion years or so until your doubt disappears; doing so will not affect the argument.) At some point there is evidence that everybody would accept for biological activity (even if we have to move far ahead in time to, say, only 2 billion years ago).

The environment on Earth clearly underwent a transition from a time in which only geochemical processes were operating to one in which biochemical processes were operating as well. Was there a distinct moment when Earth went from having no life to having life, as if a switch were flipped? The answer is "probably not." We think that the origin of life involved a transition through an "RNA world." In such a world, molecules composed of a string of nucleic acids much like a modern RNA molecule would have been able to carry out the functions both of carrying genetic information and of catalyzing chemical reactions. That is, one RNA-like molecule could have carried out the functions that today are performed by both DNA and proteins. We can postulate the events behind both the transition from the "geochemical world" to the RNA world and that from the RNA world to the present DNA world.

The formation of RNA-like molecules has been demonstrated in the laboratory using minerals as a catalytic surface, and it has been hypothesized as possibly taking place on a surface composed of layers of organic molecules. Longer and longer chains could have been built up, with each having some catalytic capability, until one became long enough and complex enough that it was capable of catalyzing its own reproduction independent of the mineral surface catalyst. While this change may sound like a switching on of life, the crossing of a boundary, this is not necessarily so. For example, the fidelity of copying could have improved as longer chains were formed. Or a short chain might have reproduced for only a few generations before using up the chemical ability to induce additions of new bases at one end of a copied chain, while a longer chain might have gone on for many generations before petering out; it might be only a matter of semantics as to how many generations of reproduction need to occur before one thinks of a molecule as living.

In this scenario, there is no sharp boundary between living and nonliving at the time of the origin of life. The extremes are easy to identify and characterize—early on with no life, and later on with life—but the intermediate stages cannot be uniquely or readily categorized. In other words, there was no point during the transition when life clearly first existed.

A Gradational Boundary for Life: Definition by Consensus

In the same sense, the present-day boundary between living and non-living may be inherently impossible to define. There may be intermediate forms, such as viruses, that cannot be placed unequivocally in either category. Just as we postulated a gradation in the origin of life, we can arrive at a gradational definition for life itself. Think of the half-dozen or so characteristics of life noted earlier in this chapter. If an entity satisfies few or none of these criteria, everybody would agree that it is nonliving. Such entities might include the wind, the atmosphere, rocks, mountains, sand dunes, and fire.

If an entity satisfies most or all of our criteria for life, then most everybody would agree that it is alive. Certainly, any example that we would universally agree to be living appears to satisfy most of these characteristics (perhaps failing to satisfy no more than one of them). This conclusion applies to single-celled organisms such as an amoeba or the *E. coli* bacterium; to macroscopic multicellular organisms such as flies, rabbits, or elephants; and to humans.

In the middle is a gray area. The more characteristics an entity satisfies, the more likely it is that we would agree that it is living—or the greater the fraction of knowledgeable people who would accept it as living. Use of this consensus concept might allow us to determine whether or not a newly discovered entity is alive. After examining the available criteria, is there a consensus as to whether it meets most or all of the characteristics of life? For those in which it doesn't, is the failure "fatal" to the idea of it being alive or only incidental?

As we will see later, much of science is done by consensus anyway, so using a consensus means to determine whether something is or is not alive actually makes sense. And this approach provides some pragmatic advice on how to interpret observations of potentially living organisms found on other planets.

The Proof Is in the Pudding: Real-World Applications

The consensus approach to determining whether something is alive is useful only if it can provide guidance in real-world situations. We will apply it to three examples: viruses, ancient terrestrial fossils, and rocks that came from Mars. We will also discuss the (at present) hypothetical case of looking for life on Mars either in situ or by examining samples collected there and brought back to Earth.

Modern Life: Are Viruses Alive?

Viruses are, in essence, extracellular nucleic acids that are encased in a protein coating. The nucleic acids can be either DNA or RNA. Viruses do not carry the machinery necessary to replicate, and so cannot reproduce independently; instead, they infect biological hosts and co-opt the host's biochemical machinery to make multiple copies of the viral DNA or RNA. The host cell bursts and releases these new copies into the environment, where they infect additional cells, create new copies, and so on.

On the one hand, viruses can be considered alive: They have the ability to replicate themselves, although an intermediate host is required. They are capable of competing via Darwinian evolution, because mutations create changes in the viral genome that give their bearers different abilities to compete and out-reproduce other viruses. And they contain RNA and/or DNA, and thereby satisfy the closest thing we have to a litmus test for terrestrial life. On the other hand, viruses are not able to reproduce without borrowing the biochemical machinery of other organisms, and thus they cannot carry out their complete life cycle alone. In this sense, they fail Gerald Joyce's definition of life because they are not "self-contained" chemical systems.

In the simplest sense, of course, no organism is entirely self-contained, because all organisms interact with their environment. Some fail Joyce's definition because they are symbiotic with other organisms and rely on them for very specific functions. Some bacteria take in chemicals that are produced as waste products by other organisms; for example, there are bacteria that live within tube worms at seafloor hydrothermal systems and provide nutrients without which the tube worms cannot survive. Some bacteria live within the gut of animals, feeding on the host's partially digested food while simultaneously helping the animal to digest the food; they are present within both humans and cows, for example. Other bacteria have become so intimately integrated with their symbiont that they no longer can live alone; chloroplasts and mitochondria, for example, which are integral parts of cells, are thought to have once been independent organisms.

Viruses occurring within each of the three major branches of life (archaea, bacteria, eukarya) share common characteristics, suggesting that they may predate the ancient split into these branches. Other viral characteristics have been interpreted as predating the origin of life. It remains to be seen whether viruses played an important role in the origin of life, originated as miscopied pieces of the DNA of ancient organisms, or represent an independent origin of life on Earth.

Although viruses are often singled out as examples of nonliving entities, the categorization is obviously not that simple. They meet some of the requirements of our definition but not others. There is no consensus in the matter, and they appear to fall in the middle of our sliding scale of life. Interestingly, there are bacteria that, like viruses, require some of the biochemical machinery of their host organisms in order to function; rickettsias and chlamydias fall within this category. And the smallest bacterial genomes are only a little bit larger than the largest viral genomes. There also are entities known as viroids that consist essentially of naked RNA molecules. Viroids are simpler than viruses because they lack the protein coat and some of the enzymes that often accompany the DNA and RNA in the viral interior. They also are capable of infecting organisms and co-opting the biochemical machinery of the host to carry out all aspects of their reproduction. From viroid to virus to small bacteria to larger bacteria: Does the sequence represent a gradational transition from nonliving to living?

Do Ancient Terrestrial Fossils Contain Evidence of Life?

Work done during the 1980s and 1990s, led in large part by Bill Schopf, a paleontologist from UCLA, pushed back the oldest fossil evidence for life to about 3.5 billion years ago. The evidence took the form of fossil entities that looked like fossilized single-celled organisms. Their appearance was not overwhelmingly convincing by itself, however, because the fossils had been severely degraded due to the heat, pressure, and other geological processes that took place during the billions of years since they formed. These entities formed a continuous "series" in terms of their physical characteristics that seemed to fall in place with entities in younger rocks that looked much more like fossils. That is, the younger ones were convincing, and the older ones looked enough like the younger ones given their substantial degradation to conclude that they were biological as well.

Recent reanalysis of these same fossils, however, has called their biological nature into question. It turns out that the fossils are not from rocks dated at 3.5 billion years but in a cross-cutting rock layer that is younger by some unknown amount. In addition, some of the characteristics of the putative fossils do not match up cleanly with those that would be expected of living things. Some have branching extensions, for example, that are not found in the biological world, or are at one end of a continuum of characteristics (size, shape) whose other end is not thought to be biological. In addition, laboratory experiments have demonstrated that similar-looking features can be formed by nonbiological processes.

Are these features biological or not? Right now, there is no consensus in the scientific community, which has responded to the new information by trying to get more data that might shed light on the issue. Without a consensus, and without more data that would lead to one, it is hard to believe the proponents of either extreme viewpoint. One group says that these features are clearly biological; the other says that they are clearly nonbiological. In this instance, the entities in question either were or were not alive; the only issue is our ability to decipher the evidence. The scientific process here is following our view of a "consensus" definition of life—as more data are obtained, a consensus regarding the best interpretation of these features will ultimately develop.

Does a Meteorite from Mars Contain Fossilized Life?

To date, more than thirty meteorites thought to have come from Mars—based on their age, oxygen isotope composition, and the similarity of trapped gases to the composition of the martian atmosphere—have been found on Earth. One of them, known as ALH84001 and collected in Antarctica in 1984, has been a controversial linchpin in the discussion of possible martian life. This rock is 4.5 billion years old, which means that it was present on Mars at about the time the planet formed, and was present as well when the surface environment might have been more conducive to life than it is today. Deposits of carbonate minerals fill voids within the rock. Such compounds are formed when liquid water flows through rock and dissolves minerals that then precipitate elsewhere within the rock—processes that occur very commonly here on Earth. This evidence of the occurrence of liquid water on the surface of Mars made ALH84001 an interesting rock in which to look for evidence of life.

A group led by David McKay and Everett Gibson at NASA's Johnson Space Center examined ALH84001 carefully for evidence of fossilized life within it, and published their results in 1996. They identified several characteristics that they thought were consistent with the presence of life. These included: (1) the presence of a specific type of organic molecules that could be either decay products from living organisms or precursors of life, (2) the presence of minerals within the voids that were adjacent to each other yet out of chemical equilibrium with each other, (3) the presence of magnetite grains with sizes and shapes very similar to those thought to be produced only by bacteria here on Earth, and (4) the presence of morphological shapes reminiscent of terrestrial microbes. While none of these characteristics could be attributed uniquely to biological activity, the group argued that, taken together, they made a strong case for it.

The announcement of evidence indicating that Mars might once have had life created a major public stir and triggered intense scientific analysis of the samples. Over the next six years a couple of dozen scientific teams examined ALH84001 in more detail and were able to provide a much clearer view of its history and characteristics. Today, each of the original characteristics once believed to indicate fossilized life has been called into question as either an artifact of the handling, processing, and analysis or as something that could be produced by nonbiological processes. The discovery of modern terrestrial microbes contaminating the rock, for example, has cast doubt on the extraterrestrial origin of any biological features.

In this case, the consensus that has emerged is that none of the features identified in the rock requires the existence of martian biota, and that all of them are likely explainable by nonbiological processes. This conclusion is not unanimous; many respectable scientists have come down on the "biology" side of the features. But it is clear that any conclusion that this rock contains evidence for martian biota is premature. We need more evidence before it will be possible to conclude that these features were formed by biological entities. New evidence continually comes into play in science, of course, and the nature of any consensus can change over time.

Searching for Life Elsewhere

If we have trouble identifying evidence of living organisms in terrestrial rocks even after more than thirty years of analysis, and if we can be so misled by analyses of extraterrestrial rocks, how are we ever going to identify life elsewhere with any certainty? What would be sufficient or convincing evidence?

One view is that a search for life on Mars, for instance, has two possible outcomes. We might identify features that are immediately accepted by knowledgeable scientists as convincing evidence for martian life. This evidence could take the form of obvious cells with easily identifiable cell membranes and/or interior structures, cells caught in the act of multiplying, and so on. Such obvious cells could be either living (extant life) or fossilized (past life). A second possibility is that the evidence might be either less compelling or more ambiguous, engendering years of debate and uncertainty such as occurred in the debate over ALH84001. Over time, a consensus might or might not develop, but, as with ALH84001, even a consensus probably would not be unanimous and would be accompanied by never-ending debate over whether life was indeed present. A third outcome is possible, of course: little or no evidence indicative of life might be found. In this case, an immediate consensus might even form that there was no life in the sample, but this result is different from what would be required to reach a conclusion that there was no life on Mars.

A second view is that it is not possible to predict what we will find or how we will react to specific discoveries. Finding entities within a martian sample that meet some of the criteria for life would not necessarily be convincing evidence for the presence of life. Chemical or mineralogical structures and order are not necessarily proof of life, but their presence would point to places worth examining in more detail. Further examination might identify features that could point in a convincing way to either a biological or a nonbiological origin for them. This approach, championed by Ken Nealson of the University of Southern California among others, suggests that we cannot actually know in advance what it might take to convince us that we have found life, but that we may know it when we see it; at the least, we can hope to see something that would require additional investigation.

Where Does That Leave Us?

Do we need a definition of life? Does such a definition help us in understanding the origin and evolution of life on Earth or the environmental conditions required to support life? Does it help us in determining whether there is life elsewhere?

It seems clear that we do not yet have enough information to arrive at a unique definition of life. At the same time, attempting to define life focuses attention on the need to understand life's characteristics. And it will be of help as we explore the ancient fossil record on Earth and the fossil or possibly biological record on other planets. Without a unique definition, though, it will be only by consensus within the knowledgeable scientific community that we are able to reach any conclusions.

Or, we can always fall back on Mark Twain's definition: "Life is just one damn thing after another."

3

Is Astrobiology Science?

WE HAVE NO EXAMPLES OF LIFE TODAY OTHER THAN
that on Earth. For that reason, astrobiology and its close cousin exobiology
are often described in a somewhat derogatory way as scientific disciplines
without a subject. Is this a valid criticism or just a glib joke?

Before we can ask whether something is a science we must first ask
two more fundamental questions: What constitutes science? Where is the
boundary between science and nonscience? The answers to these questions
lie more in the realm of philosophy of science than in science itself. Unfortunately, relatively few scientists seem to be interested in the philosophy of
science beyond a superficial level, even though its goal is to describe how science operates. Most scientists, if they think about it at all, seem to hold the
view that they are too busy doing science to pay attention to philosophy of
science. In any case, scientists approach astrobiology from one perspective,
and philosophers of science approach it from another. If the two sides can
be brought into a dialogue, we can learn something about both disciplines.
There are several reasons why such a dialogue is important.

First, the scientific method as commonly understood by both scientists
and nonscientists does not adequately describe how much of astrobiology
functions. A particularly important aspect of astrobiology is our desire to
explore the universe to find out what is there. This "exploration science"
is fundamentally different from the experimental or hypothesis-driven sci-

ence that is typically described by the scientific method. Thus, as scientists, astrobiologists have difficulty explaining to others how we do our science or even why astrobiology is science. This issue does not come up very often, because astrobiologists seldom discuss or publish their most detailed results for the larger scientific community. We usually spend our time with people who have been trained to understand, or to at least accept, what we are doing.

Second, astrobiology is not a "unified" science. There are different ways of "doing" science that apply to the different fields within astrobiology. As these involve different ways of asking and answering questions, practitioners in one area often find it difficult to understand what those in another area are doing, how they are doing it, and why they are taking a particular approach. For example, atmospheric and planetary dynamics are usually predictive sciences in which one can do the "forward" calculation of how a particular system will behave under a certain set of assumptions and compare the calculations with the characteristics of a real system in order to draw conclusions about the nature of the system. In contrast, planetary geochemistry usually is seen as an experimental science. A researcher takes a sample in the laboratory, for example, and analyzes it to measure some characteristic or property, then draws conclusions by comparing the measurements to some theory or with equivalent measurements of a system on which known processes have acted. Planetary geology operates by observing the morphology (or the physical properties or the composition) of a surface and inferring by analogy with the Earth what processes in the past made it that way. Each of these subdisciplines works in fundamentally different ways. Scientists trained in one discipline may understand intuitively how their own field works but may not appreciate the subtleties of another.

Much of the cutting-edge research within planetary science and, more broadly, astrobiology works at the boundaries between the traditional disciplines. It is important for the scientists in one field to understand how the neighboring fields work, to prevent misunderstandings of how results from other disciplines were obtained and should be interpreted as well as to prevent blind and sometimes inappropriate acceptance of conclusions drawn from the other fields. A clear understanding of how each subdiscipline works, how conclusions are drawn, and the strengths and uncertainties in the conclusions is necessary for their effective integration. Philosophy of science can play a role here because these are exactly the types of questions it addresses.

A third reason to understand astrobiology from the viewpoint of philosophy of science is that much of astrobiology involves evaluating and comparing competing hypotheses. It is important that astrobiologists understand the nature of the competing hypotheses and how best to choose among them.

These issues lead to the underlying issue of how we do science, both as individuals and as a scientific community. This more basic question can be addressed through a discussion of how astrobiology is done and how it relates to the intellectual threads identified in philosophy of science. While the following discussion may seem naive and oversimplified to a trained philosopher of science, it will address some of the pertinent issues from the perspective of the practicing scientist. As such, it begins to identify some of the issues that will arise in any dialogue between science and philosophy of science.

The Concepts of Confirmation and Falsification

As the modern concept of science began to develop a few hundred years ago, a method of doing it evolved that entailed making systematic observations in order to evaluate ideas. Thus, "doing science" involved developing hypotheses, making observations in nature or performing experiments in the laboratory to test the hypotheses, and using the results to confirm or refute them. Edmund Halley's prediction of the return of a comet is a good example of how the idea of confirmation was applied.

Isaac Newton published his theory of mechanics in 1687, giving astronomers—through Newton's law of gravity—the ability to calculate how planets and comets orbited the Sun. Halley had seen the comet that would later carry his name in 1682. Using his observations of the location of the comet in the sky on many different nights as it moved along its orbit he determined the shape and orientation of the ellipse traveled by the comet and the velocity at each point on this ellipse. He then used this information to predict where in the orbit it would be at any instant and where it should appear in the sky. He was able to determine that the comet's orbit had a long period, but his observational techniques were not good enough to allow him to determine whether the orbit was elliptical or parabolic. If parabolic, the comet would not be bound permanently to the Sun but would represent an object making a one-time pass through the solar system; if elliptical, it would be in permanent orbit around the Sun and presumably would come around again and again.

Halley compared the orbit of the 1682 comet with the orbits of two dozen other comets. These comets were the only ones for which there were sufficient observations to allow approximate determinations of their orbits. He found that two other comets had orbits that closely matched those of the 1682 comet: the comets of 1606–1607 and 1530–1531. He concluded that it was unlikely that three different comets had orbits so similar to each other, and that they probably represented three appearances of the same comet as it orbited the Sun. This conclusion allowed the further conclusions that the comet was bound to the Sun in an elliptical orbit, that it had a seventy-five-year period, and that it should reappear around Christmas 1758. Although Halley was no longer alive when it returned, it did return, and the comet was named in his honor.

The return of Halley's comet was deemed to be a brilliant confirmation of Newtonian mechanics. In particular, it appeared to confirm (1) that Newtonian mechanics governs the movement of objects in our solar system, even of objects other than planets; and (2) that, by virtue of their orbiting the Sun, comets are an integral part of our solar system.

Let us apply this example to the general concept of confirmation in science. Halley's observations were found to be completely consistent with the consequences of the hypothesis of gravity that Newton had put forward. Does this mean that Newton's idea was correct? How many confirming observations such as this would be necessary to establish the truth of a hypothesis? Could Halley's observations have been consistent with other hypotheses about how the solar system operated, even ones that had not yet been thought of? Does this confirmation tell us something significant about all comets, including ones that have not yet been discovered?

In the early 1900s, Karl Popper recognized a problem in this application of the idea of confirmation. No matter how many confirming observations might be made about a theory, one could never prove a hypothesis to be correct. It would take only one observation truly inconsistent with the observations to disprove it. He suggested instead that, while a hypothesis can never be confirmed as "true," no matter how many observations are made, it can be disconfirmed or falsified by new observations.

In the case of Newtonian mechanics, the determination of the orbits of the planets and of Halley's comet provided excellent confirmation. We know today, however, that some later observations were inconsistent with Newtonian mechanics, specifically the precession of the orbit of Mercury and the bending of light by a massive object such as the Sun. As a result, Newtonian gravity has been superseded by Einstein's theory of gravity that includes relativity. While one could argue that it doesn't make that much difference, and that Newtonian mechanics is "good enough" in most cases, it certainly makes a difference if we are sending a spacecraft to another planet. Despite the many confirming observations, Newtonian mechanics was eventually discarded.

Let us examine the idea of confirmation using the approach of formal logic. Suppose we have a hypothesis, H, and an observation that we think is a test of the hypothesis, T. Applied to Halley's comet, H is the concept of Newtonian gravity and its consequences as they relate to planetary and cometary orbits, and T is the observations by Halley of the comet of 1682, combined with the observations of the orbits of two dozen other comets.

The concept of confirmation takes the following form: If H is correct, then T follows. That is, if Newtonian gravity is correct, then the observations of the comet will be consistent with an object orbiting the Sun, and it would follow that the comets of 1606 and 1530 likely were the same object. The observations tell us that the test condition T is satisfied. Therefore, the conclusion is that H is confirmed—Newtonian mechanics is correct. (Actually, the logic is such that H need not be true for T to be true, so the conclusion should be that the hypothesis H is not refuted or falsified.)

We can set up cases that have similar logic but for which confirmation is not appropriate. That is, there are examples in which H can be wrong but T can still be true. For example, suppose that H posits that because life originated rapidly on Earth, it should have originated on Mars as well. The consequence of H being true would be that there is life on Mars (or was at some past time). If we find life on Mars (T), it would suggest that our hypothesis is confirmed and that Mars had an independent origin of life (H). However, there are other ways in which life can exist on Mars; for example, it could have been transferred there from the Earth on a meteorite ejected into space by a large impact (in fact, that is a viable hypothesis). Thus, while the test condition may be consistent with the hypothesis, it does not necessarily confirm the hypothesis.

On the other hand, finding no evidence for past or present life on Mars would suggest that life never originated independently there. (This conclusion would follow only if we searched far and wide on Mars for evidence, in all of the places in which life could have existed, and found none. Even then, it follows only under the additional assumption that, had life been present, it would have left behind a record of its existence.) Such is the case whether or not life was transferred there from the Earth.

Thus, it takes only one failed test to falsify the hypothesis. In Popper's view, this was the key to science. If we have multiple hypotheses that might explain our observations, we should test each of them, throw out the ones that don't work, and keep the ones that do. By this process we can eventually identify and discard incorrect hypotheses. Falsified hypotheses can be modified to try to make them more consistent with the observations. The modified hypotheses that we retain represent increasingly better representations of how the world actually operates, in that they have been tested under an increasingly wider variety of conditions.

This idea of falsification has been a key part of most scientists' views of science for nearly a century, and it may be the view most widely held today by scientists. But it has its own problems. In the real world, no observation can be made and no experiment can be designed that tests only a single condition. Rather, each test that we perform is based as well on a large number of assumptions that we normally don't think about. In the example of Halley's comet, for instance, these assumptions include the idea that the light we observe tells us the true directions to objects, that the Earth's rotation axis doesn't change from one day to the next, that Earth's orbit around the Sun doesn't change from one day to the next, and that the object that we see on one night is the same object that we see in a similar location in the sky on another night. More subtle is the assumption, for example, that comets don't come from only a few locations in the sky, in which case the comets of 1530, 1606, and 1682 conceivably could be different objects that had very similar orbits; or that no forces other than gravity can act on the comet (this one is known to be false, in that the gases subliming from a comet's surface impart a force to the comet). If the test condition is not satisfied, then either the original hypothesis can be rejected or one or more of these auxiliary assumptions should be rejected. That is, if both the hypothesis and the auxiliary information are correct, then the test condition is satisfied. If the condition is not satisfied, then either the hypothesis is incorrect or one or more of the auxiliary assumptions is incorrect, but it may not be possible to determine which. In the example of Halley's comet, none of the auxiliary assumptions listed is so incorrect as to invalidate the conclusions. We can easily identify any number of other experiments that were not sufficiently accurate or precise or that were carried out poorly as examples in which falsifying the auxiliary assumptions is more appropriate than falsifying the main hypothesis.

Let us pick a more interesting example that also shows the importance of the auxiliary assumptions a little more clearly: the Viking spacecraft missions to Mars in 1976 and their search for life there. The two Viking landers were searching specifically for signs of metabolism within the martian soil, and each performed three experiments testing for different indicators that metabolism might be occurring. The first experiment looked for uptake into the soil of carbon, which was supplied in the form of gaseous CO and CO_2. The second experiment looked for gases that would be released by the utilization of nutrients that were supplied to the soil. The third looked for release of carbon back into the atmosphere as a by-product of the metabolism of nutrients that were added to the soil.

The initial results of all three experiments were what had been predicted to occur if life was actually present—the uptake or release of the expected gases. A second run was done in each case, however, in which the sample was first heated to temperatures high enough to destroy any organisms similar to terrestrial life. In the first two experiments, the results were the same as those in the initial runs, which suggested that some nonbiological (i.e., geochemical) reaction was taking place. In the third experiment, the heat severely degraded or eliminated the response to nutrients, which was the response that had been predicted to occur if living organisms were present in the soil. Did the third experiment, the "Labeled Release" experiment, actually find martian life? Or were auxiliary assumptions playing an important role here?

A fourth experiment was relevant as well, even though it did not test specifically for biological activity. A gas chromatograph/mass spectrometer (GCMS) was designed to detect organic molecules in the soil. Organics should have been present even in the absence of biological activity, brought there by incoming meteorites and asteroids. In the GCMS experiment, a soil sample was heated enough to vaporize any organic molecules that it contained. These molecules then passed through the instrument, where they were identified by their masses. The composition of the organic molecules should indicate their origin—meteoritic or biological. No organic molecules were detected at about the parts-per-billion level, meaning that if active or dead organisms were present in the martian soil, it was at extremely low levels.

Furthermore, the results of the first two biology experiments could be explained in terms of nonbiological reactions between the gases and nutrients supplied by the instruments and superoxides in the soil. These superoxides would include molecules such as hydrogen peroxide (H_2O_2) that might be created in the atmosphere by chemical reactions and then diffuse into the soil. The presence of superoxides would also explain the absence of meteoritic organics, as organics could react chemically with the superoxides and be broken apart into CO_2 and H_2O.

There were not enough organic molecules in the soil (given the parts-per-billion upper limit) to allow the results of the Labeled Release (LR) experiment to be considered biological. The same type of "volatile" reactions that were suggested by the other two biology experiments could also explain the LR results. Many of the LR results, specifically the very rapid initial response and the lack of a follow-up response to a second addition of nutrients, also were not entirely consistent with a biological response. Thus, the best interpretation of the total suite of results was that the experiments had not detected any form of life, despite the fact that the results met the prelaunch criteria for detection of life.

The hypothesis, H, in this case is that life is present on Mars and will produce a response such as that seen in the LR experiment; the test condition, T, is the apparent biological response of the experiment. Following the preestablished criteria for life, one could interpret the LR results as confirming II. This situation exposes one of the problems with confirmation: There is at least one alternative hypothesis—the presence of superoxides, based on the absence of organics—that would also be consistent with the results. Thus, confirmation is not equivalent to proof.

Recent analyses by Steve Benner and his colleagues suggest that organic molecules could in fact have been present in the soils sampled by Viking in a reduced form that is less volatile than normal, and that the GCMS experiment might not have been able to detect them. Where does that leave us? It is not at all clear what the best interpretation is of all of the data. Although active biology is not ruled out, it does seem much less likely than the alternative hypotheses.

A Change in Our Paradigm of How Science Works

In the 1960s, Thomas Kuhn, a historian and philosopher of science with a background in physics, recognized that the concept of falsification, while useful, does not actually describe how major breakthroughs in science occur. Much of what scientists do involves confirmation and falsification, but Kuhn noticed occasions in which the fundamental intellectual concept of how something worked changed completely. The changes were so dramatic that they could not be seen as simple modifications of an existing hypothesis but instead represented a complete rejection of an existing hypothesis and the introduction and acceptance of a completely new one. He labeled these changes "paradigm shifts."

Astronomers' discovery that the Earth orbits the Sun rather than vice versa is an excellent example of a complete change in perspective. In the time of Aristotle, in the fourth century BC, the Earth was thought to be at the center of the universe and the Sun, Moon, and planets were thought to move around it on circular orbits. This idea stemmed from the appearance of the nighttime sky as an inverted bowl surrounding the Earth, combined with the prevailing concept that the heavens must be ideal in their nature, and nothing could be more ideal than circles and circular motion. Simple circular motion could not be the only movement of the planets, however, as some of them were occasionally seen to move backward in the sky with respect to the apparently immobile stars. This retrograde motion could be explained if the planets actually moved on smaller circles that were centered on a point that itself orbited the Sun on bigger circles. If the planets moved within these "epicycles," their motions would still be circular and the Earth could still be at the center of all motion. Small deviations of the resulting motion from the observed motion of the planets could be resolved by adding additional epicycles. Seven hundred years later, the astronomer Ptolemy developed this idea in detail and made specific predictions as to where the planets should appear in the sky; these predictions were used for more than a thousand years.

By the time of Copernicus in the mid-1500s, however, deviations from these predictions had made Ptolemy's calculations increasingly inconsistent with the observations. More and more epicycles were needed to bring the predictions back into alignment. In trying to reform the Ptolemaic calendar, Copernicus proposed a simpler system, one in which the Sun was at the center of all motion instead of the Earth.

One hundred fifty years later, the discoveries by Galileo and Kepler, the theory of Newton, and the confirmation from Halley's comet all supported this shift in perspective. A complete change in mind-set removed the Earth from the center of the universe. This change represented a fundamental shift in how humans viewed the universe. Once it was recognized that the Earth orbits the Sun on an elliptical orbit, the old theories and calculations were discarded. All subsequent observations and analyses were undertaken on the basis of, and interpreted in light of, the new perspective.

Paradigm shifts occur in modern times as well, of course. A recent example is the revolution in our understanding of the motions of the surface and interior of the Earth embodied in plate tectonics. Beginning in the 1950s, new observations suggested that the Earth's crust consists of numerous separate plates capable of moving independently. Motions of the crust over most of the Earth's surface appeared to occur primarily at the seams between large plates. Mountain-building processes occurred at these same plate boundaries, and both local- and global-scale deformation and movement seemed related to these motions. The surface movements, and the creation of new crust and destruction of old crust, were seen as the manifestations at the Earth's surface of processes that were occurring in its interior. Once one understands the global-scale behavior of the Earth and how local events fit into the broader scheme of things, it is not possible to go back to a system in which events are local and unconnected. Much as with biology and evolution, nothing in the geology, geophysics, or geochemistry of the Earth makes sense except in the context of plate tectonics.

In Kuhn's view of how science works, scientists generally operate within a regime that he called "normal science." During normal science, scientists make measurements within the context of the existing paradigm in their discipline in order to try to understand relatively narrow problems. Most scientists do this type of "puzzle solving" for their entire careers. When a hypothesis needs to be modified, it is usually done in small increments, and subsequent research again aims at confirmation and falsification or exploring the ramifications of the existing paradigm. Occasionally, however, measurements or observations that are inconsistent with the prevailing paradigm crop up, and it becomes increasingly difficult to modify the hypothesis to bring it back into alignment. Often, so many modifications are required, one on top of another, that the hypothesis becomes too cumbersome to be viable. Kuhn referred to this kind of occurrence as a time of crisis. At such times the old theories can no longer offer good solutions to the problems they were designed to explain. New hypotheses might be generated and explored, and some might be rejected via falsification. Some scientists might accept one new hypothesis while others accept another. But there is no consensus as to which new hypothesis might be best or what the implications are. If one of the new hypotheses becomes widely accepted, a paradigm shift has occurred. This new idea becomes the accepted or standard wisdom in the field and the baseline against which other ideas are compared.

In our example of the Earth-centered universe, it was the increasing disparities between the predictions of planet locations made by assuming circular orbits and a limited number of epicycles, combined with the cumbersome and completely unphysical model of adding epicycles upon epicycles, that caused the crisis. Copernicus was the first to face the crisis head-on and suggest an alternative approach, but more than a century passed before the paradigm of a Sun-centered universe became generally accepted.

What Is the Role of New Observations in Astrobiology?

Popper's and Kuhn's views of how science works apply quite well to some aspects of astrobiology. One example that fits within this context involves deciphering the earliest geological history of the Earth. Ideas are developed and then tested using field and laboratory observations. Hypotheses are modified, and new field and laboratory measurements and observations are defined that might be able to provide more details. Much of the development of ideas follows the confirmation and falsification approach, in which the hypotheses that survive are the ones that fit best with the observations. Instances of paradigm shift as described by Kuhn occur as well; for example, dealing with new discoveries pertaining to the nature of the environment and climate on the earliest Earth.

On the other hand, some of astrobiology's component disciplines do not fit within these models of science. Many disciplines progress largely through measurements and/or observations that are so profoundly new that they immediately change our fundamental view of the problems. Within planetary science, for example, major advances have come about each time a new spacecraft has been sent to a planet. The measurements these spacecraft have made have opened up entirely new windows into the nature of the planet. More often than not, the new data have made most or all of the existing hypotheses about how the planet developed obsolete and new hypotheses have had to be generated.

An example of this process comes from the first Voyager spacecraft's passage through the Jupiter system in 1979, during which the satellites of Jupiter were observed close up for the first time. Io, Europa, Ganymede, and Callisto, each comparable in size to or larger than our own Moon, went from being points of light in orbits around Jupiter to being worlds in themselves.

Prior to the Voyager flyby, we knew from reflectance observations that Europa is covered with water ice and that the ice has relatively few contaminants; we knew that its orbit is locked tidally to the orbits of Io and Ganymede; and we knew that its orbit is not quite circular and that as a result there should be some interior heating resulting from dissipation of tidal energy. The Voyager flyby allowed us to learn specific details about the geology and geophysics of Europa for the first time. We found out that the surface has very few craters and must be geologically active; what the satellite's size and average density (and thereby the water ice content) are; and what geological processes have taken place on the surface.

In essence, we went from a situation in which we had no strongly constrained paradigm that could describe the current nature of Europa, the processes that controlled its evolution, and its current state, to one in which we could develop such a paradigm based on the new observations. The major results did not come from clever insights into what might have happened on Europa or from an ability to make specific new observations to test an existing hypothesis. Instead, they come from sending a spacecraft to the planet with a variety of instruments and seeing for the first time what the surface looks like and what is in the atmosphere.

A second example comes from the recent discovery of planets orbiting other stars; more than 150 have been identified to date. These planets are detected by observing the planet's gravitational effects on the parent star. The planet and the star each orbit around their mutual center of gravity, and we can observe the apparent wobble of the star back and forth as it moves around this point, and from this wobble infer the existence of a planet. Observations of the wobble can tell us both the mass of the planet and the shape of its orbit (technically, this yields the product of the mass and the sine of the orbital inclination). This technique is sensitive enough to detect planets with a mass as low as that of Neptune.

The inferred shapes of these newly discovered planets' orbits and their distances from their parent stars were completely unexpected. Most of the planets that have been discovered either orbit very close to their star or are in very eccentric orbits. Very few of the newly discovered solar systems look anything like ours, with its Jupiter-like planets relatively far from the Sun and in nearly circular orbits. Our preexisting paradigm of how planets form and evolve was based on detailed observations of our own solar system. The new observations of other planetary systems represented the first influx of real information and immediately superseded the old understanding. New hypotheses of planet formation and evolution had to be developed essentially from scratch. Within several years, a half-dozen explanations for how planets could come to exist either so close to their star or in such eccentric orbits emerged; some have since been rejected, but most are still awaiting tests by new observations or theoretical developments.

Although there were existing paradigms for Europa and for planetary formation, in neither case was there a standard Kuhnian crisis as new observations were either forced to fit into the existing theory or did not fit at all. Likewise there was no development of a new paradigm to compete with the old, and no transition from acceptance of the old paradigm to acceptance of the new one. Instead, the development of a new technology allowed a new type of observation to be made. This was followed by a dramatic and sudden influx of observations and measurements that made all existing data obsolete and irrelevant, and these data provided a basis for the equivalent of the development of an initial paradigm.

The field of planetary science is based in large part on this type of discovery-driven or "exploration" science. We have been sending spacecraft to other planets in our own solar system since the 1960s. The development of our understanding of how the planets work has been so intertwined with the results obtained by these spacecraft that the two cannot be separated to any significant degree.

This idea of exploration science can even be seen as a continuing pattern using the example of the planet Mars, which has been visited by spacecraft many times since 1965. *Mariner 4* flew by in 1965, *Mariner 6* and *Mariner 7* flew by in 1969, *Mariner 9* orbited in 1971, the Viking mission consisted of two landers and two orbiters in 1976, the Russian *Phobos* spacecraft operated briefly in orbit in 1989, *Mars Pathfinder* landed in 1997, *Mars Global Surveyor* orbited beginning in 1997, *Mars Odyssey* orbited beginning in 2001, the European *Mars Express* orbited beginning in 2003, and the two Mars Exploration Rovers landed in early 2004. Almost all of these spacecraft provided unique data sets different from all preexisting data sets and with much more detail, and each opened the way for the creation of a new paradigm for the evolution of Mars. While the instruments that flew on each spacecraft were often proposed and designed to test specific hypotheses, the real value of the results often came not from these tests but rather from the fact that observations of a specific type were being made for the first time. It is these anticipated yet unexpected results that cast the approach as exploration science rather than the more traditional hypothesis-driven science.

Of course, this situation could be seen as an extreme example of how a Kuhnian paradigm shift works. New measurements put old hypotheses into doubt, and new ideas come up to replace them. It happened so quickly in these examples, however, that the process should be recognized as a fundamentally different one. In this type of process, new hypotheses are developed, evaluated, and cast aside within a matter of days. Over a period of days, months, or a year, a number of different ideas might each take their turn being the standard wisdom and being rejected in turn. Only over a period of perhaps one or more years does the discipline settle down into a single consensus view of the planet. This consensus develops during the time that the new data are being assimilated, and it may last only until the next mission, at which time the process starts over again.

Science Today

There appears to be no single model of how science operates today that works in all situations. Although many scientists point to confirmation and falsification as the key elements of how science is done, each of these has problems. Confirmation cannot establish the truth of a hypothesis because no amount of observations or measurements can demonstrate that a hypothesis is correct or true; it can only show that a hypothesis is valid in specific instances. Even a confirmation based on the most stringent test that can be devised cannot necessarily be generalized to all instances.

Falsification has no such problems. A single failure to confirm a hypothesis is sufficient to demonstrate that it is false. Because auxiliary assumptions exist that can be falsified instead of the main hypothesis, however, falsification also is not strictly valid. And because science often involves two or more competing hypotheses without having the data in hand to allow a valid choice, falsification does not necessarily leave us with only a single viable hypothesis.

The concept of a paradigm shift, by its very nature, falls outside the concepts of falsification and confirmation, and represents more than just an incremental advance in our understanding of a problem. At the same time, the concept tends to be overused by some scientists who inflate the importance of their discoveries. If textbooks were being rewritten as often as scientists say that their results should cause them to be, we would never be able to print any books! Nevertheless, there are real examples of paradigm shifts in astrobiology that represent the true development of new paradigms. And as the ongoing development of new technology—including new laboratory techniques and instruments, orbital telescopes, and interplanetary spacecraft—literally opens up new worlds to view, there will almost certainly be more.

It should by now be apparent that no single approach can describe how astrobiology (or any science) works. Even carefully chosen examples cannot quite be made to fit within the confines of any one of these perspectives, although each example I have offered brings out salient points about how different aspects of science work. The overall nature of science must involve parts of each of these approaches to differing degrees.

Even if we cannot choose one model of science over another, we can use the differences between these approaches to ask some fundamental questions about the nature of science and how we understand it to work today.

Is There an Absolute Truth in Science?
In Popper's view, hypotheses are modified (or discarded completely) in response to observations that are not consistent with them. Over time, in-

correct hypotheses disappear, and those that remain are increasingly accurate. In this view, science is objective, in that an observation either is or is not consistent with the pertinent hypothesis, and the hypothesis is either falsified and must be either modified or discarded, or it is confirmed and can stay on the table. The hypotheses that are still viable become increasingly better tuned to the phenomena they describe, and they become increasingly better approximations to the truth. This view implicitly suggests that there is an absolute truth in science and that our view of the world continually approaches that absolute truth.

Kuhn's ideas about paradigm shifts appear in sharp contrast to this view. While he recognizes that small modifications can be made to hypotheses along the lines that Popper suggested, he focuses on the larger change that occurs when a new paradigm replaces an old one. While the smaller changes in a hypothesis seem to give better and better approximations to the truth, the potential for there to be sudden shifts in perspective means that we cannot be sure that our fundamental picture will not change. Thus, we might find ourselves moving closer to a particular approximation of the truth, and then suddenly find that this approximation is really not very good at all and that we are suddenly, from this new perspective, far away from a good approximation. While each new paradigm might seem to be closer to the truth, we can never be sure that another new paradigm isn't lurking just around the corner. That is, we can never know if we are close to the truth, because we have no way of knowing what the truth or reality is.

One example of such a change is the shift to a Sun-centered universe, which involved accepting Newton's ideas about gravity. Discrepancies that were discovered later in the location of the planet Uranus did not cause astronomers to reject Newtonian gravity but instead to postulate the existence of another planet that also was exerting a gravitational influence on it. The subsequent discovery of Neptune and the determination that it was responsible for the observed perturbations in Uranus's orbit were further confirmation of the veracity and applicability of Newton's law of gravity. But a different set of discrepancies between predictions and observations of a planet's location—this time Mercury—helped astronomers to confirm Einsteinian relativity and gravity and to reject Newtonian gravity. Might there be another version of gravity that has effects more subtle than Einstein's version?

While the universe logically can have had only one history and one set of processes that operated in certain ways, our ability to infer what that history and those processes are is limited. We need to keep separate, at least in our minds, the concepts of the underlying reality of the universe, our depiction of it based on observations and inferences, and the causal explanations that we put forward to understand it. Given the differences inherent here, we may never be confident that we really are approaching an absolute truth.

Is Science Objective or Subjective?

In Popper's estimation, science is objective in that there is one truth and we are edging closer and closer to it; hypotheses are confirmed, falsified, or modified. Two important aspects of the Kuhnian perspective force us to re-think this concept of objectivity.

First, it should be clear that, under Kuhn's view of paradigm shifts, our interpretation of observations or measurements depends very strongly on the perspective from which we are viewing the system. In our example of the dynamics of the Earth's surface and interior, our view of the driving mechanisms behind the history of the surface and interior depended on whether or not we viewed plate tectonics as responsible for movements of the Earth's crust. Prior to plate tectonics, the geological evolution of the surface was thought to depend on local events and forces (although these may not have been well understood). Under the doctrine of plate tectonics, it depends on the relationship of local properties to the surrounding geology and to the global distribution of plates, plate motions, and related effects. The nature of the cause-and-effect relationship that we infer depends on our perspective of the important processes.

Second, no set of measurements or observations can be perfect or will agree perfectly with a given hypothesis. The degree of disagreement required to be fatal to a given hypothesis is inherently subjective, as is the choice of which hypothesis among many is favored.

The upshot of both of these issues is that it is hard to imagine that there is a truly objective way in which hypotheses can be evaluated. Reasonable people can look at the relationship between hypotheses and observations in a given instance and reach different and contradictory conclusions. Neither Kuhn nor Popper tells us how to choose between competing, viable hypotheses.

This issue has important implications for how we evaluate competing hypotheses. Many scientists like to argue that science is not done by popular vote; it is based on objective analysis and interpretation. The facts that multiple interpretations are possible in any situation, that no hypothesis can be proven correct, that few if any hypotheses can truly be proven incorrect, and that well-meaning and honorable people can disagree about the evaluation of a hypothesis, however, suggest that science is in fact done to a large extent by consensus within the framework of the existing paradigm. When does a scientific community decide that a new hypothesis should become the accepted paradigm? One can argue that this happens only when one hypothesis is demonstrated as being much more likely or more plausible than its competition, and that this result is a demonstration that it is better from an objective point of view. However, "more likely" and "more plausible" are value judgments. Each person can choose where a hypothesis will fall on this scale, and a consensus is reached only when most scientists within a given discipline judge it to be the most plausible.

Is There a Cultural Influence or Bias in Science?

The current state of scientific knowledge and the broad suite of existing theories and hypotheses in a given discipline provide the context in which we interpret new results. The fact that the interpretation depends on this context may be equivalent to suggesting that culture does influence science.

Is it possible to carry this idea further and suggest that this cultural influence comes not only from the scientific culture but from the societal culture as well? Is it possible that a person's understanding or interpretation of scientific results could depend on such cultural factors as sex, race, national origin, or ethnicity? Given the influence of the broader culture on the thinking of individuals, it is easy to imagine that these factors might affect the interpretation of scientific results. Certainly we should not be too quick to reject the idea that there is a human influence on science, given that science is an endeavor done by humans and that scientists bring to their work all of the frailties, failings, and biases inherent in human nature. This idea of identifying implicit biases, even in science, is addressed by the postmodernist deconstruction movement.

Science can be considered as just another "text" that is subject to interpretation and debate. Just as one can interpret a literary work and find meaning that might differ from that intended by the author, so can one interpret a scientific work and find multiple meanings that might differ from those recognized by the experts. And, just as each individual's interpretation of a literary work can have value and legitimacy, so might each individual's interpretation of a scientific work or result. We see this concept being applied broadly in our society today—expressed, for example, in the ideas of political correctness: differing opinions are often considered equally relevant, regardless of the knowledge base and understanding on which they are based.

At one extreme, science's very results can be called into question, as the sole value of a scientific work might lie in its interpretation by the individual. Science does not necessarily have any intrinsic meaning; in fact, *science* is not easily defined. This extreme view has to be tempered, of course, with the idea that science works. Airplanes designed on the basis of scientific principles do fly, bridges don't fall down, and computers do what they are programmed to do, regardless of the observer's or user's interpretation of their value.

It seems that we must recognize the potential for culture to influence and affect science. At the same time, we must not allow this idea to carry us so far that we reject the basic conclusions that come from science.

Astrobiology as a Science, Revisited

I began this chapter by asking whether astrobiology is a legitimate science. We can now revisit this question. It would seem that the one fundamental tenet of science is that the world is inherently understandable and that we can understand how it works by observing it. We develop and evaluate hypotheses based on measurements and observations made in the real world rather than by pontificating in isolation about the way we think the world should work. This is the case whether we are examining existing hypotheses and modifying or eliminating them based on observations, developing new hypotheses that are in fundamental disagreement with existing ones, or making new observations in such a way as to supersede the previous results and force the creation of new paradigms.

Astrobiology certainly would seem to fall within this general characterization. Astrobiologists develop and evaluate hypotheses about how planets form and evolve, what makes them habitable, and how life developed on Earth and how it might interact with planets in general both in our solar system and elsewhere. We try to understand the relationship between the properties and processes in our solar system, about which we know a great deal, as well as those in the 150 or so other known planetary systems, about which we know much less.

The fact that we don't know whether life exists elsewhere is simply not relevant to the question of whether the discipline exists. Understanding the potential and actual distribution of life in the solar system and in the universe is an endeavor amenable to scientific investigation. The answers may not come within the next decade, but getting the answers is possible.

Does astrobiology have anything to gain from this discussion of the philosophy of science? How does a discussion of how science is done contribute either to the doing of the science or to determining the implications and ramifications of having done the science? The relevance of philosophy of science to the practice of science emerges in a number of the issues that have been brought out here: the difficulty of disproving or falsifying hypotheses in the real world, the lack of objective means by which to evaluate and compare competing hypotheses, the resulting idea of a cultural influence on scientific results, the ability for honest and sincere scientists to disagree about the interpretation of measurements or observations, the idea of the acceptance of scientific hypotheses by consensus, and the role of fundamentally new observations in allowing the development of brand-new hypotheses and paradigms. Especially important is the role that exploration science plays in astrobiology and an understanding of how this approach to doing science differs from the more traditional views of science.

Regardless of the approach we take, we should not lose sight of the fact that science works. The methods of science, with all of their ambiguities, conflicts, and confusion, are still the best way that we have devised to make sense of the world around us.

4

Astrobiology as a Historical Science

WHEN MOST PEOPLE, INCLUDING SCIENTISTS, THINK about the nature of science as an intellectual endeavor, they think of the view of the scientific method they learned in high school, college, and even graduate school. This view is based on a standard concept about experimental science, in which a hypothesis is formed and then tested by doing a controlled experiment. The experiment may involve setting up a laboratory activity in which the initial conditions determine the outcome, or it may involve watching an experiment that is being carried out by nature. Depending on the outcome, the hypothesis might be modified or even abandoned, a new set of predictions made based on an updated hypothesis, and a new set of experiments designed and carried out. This is the canonical view of confirmation and falsification described in chapter 3, where we also considered a number of ways in which this view does not fully describe how science works, including the paradigm shifts that were described by Kuhn and the exploration science that is an important part of astrobiology.

There is still another fundamental way in which the standard view of the scientific method does not describe astrobiology. To a large extent, the "experiments" that have been carried out by nature are so complex, with so many intertwined processes and effects, that it is impossible to separate single components of cause and effect. The outcomes of many of the processes acting on pertinent systems are inherently unpredictable because of the tremendous complexities involved. In addition, many of these "experiments" took place in the distant past. We cannot watch them occurring; we can only see their effects. For example, we cannot fully understand the processes that led to the formation of the Earth. So many processes operated simultaneously that they cannot be separated from each other, and the chaotic or random behavior of some of the processes precludes an accurate assessment of the precise way the planet coalesced.

We also may never be able to understand the exact sequence of events that led to the origin of life on Earth. We cannot determine which processes were most important, or even what specific processes necessarily operated to form life. Even if we are able to create life in a test tube today, and even if the same processes could plausibly have operated on the early Earth, they are not proof that this is how the origin of life took place on Earth four billion years ago.

Many astrobiologists face particular obstacles because they study processes that operated in the distant past. They endeavor to determine what the processes were, how they operated, and their relative importance. In this chapter we will consider the problems inherent in trying to apply the methods of experimental or predictive science in such circumstances and the validity of using the "historical" approach instead.

The Nature of Experimental, Observational, or Predictive Science

To understand how astrobiology relates to our view of how science works, we can look at examples of predictive and experimental science. Predictive science goes back to the Copernican revolution that took place from the mid-1500s to the late 1600s. Contributions by Copernicus, Galileo, Kepler, and Newton led to the general acceptance of the idea that the Earth goes around the Sun rather than the Sun around the Earth. This discovery led to the modern view of science as an intellectual endeavor; the approach based on these findings has dominated thinking about science ever since.

As I note in chapter 3, Ptolemy made his predictions of the planets' locations in the sky by assuming that the Sun and planets moved in circular orbits around the Earth, and he added epicycles as needed to account for deviations between the predicted and actual planetary positions. By the time of Copernicus, the number of epicycles required to bring the calculations into line with the planets' actual positions was so great that the system seemed too cumbersome and complex to be considered an accurate representation of reality; certainly it was too clumsy to be the product of a perfect deity. Copernicus made a seemingly small change by assuming that the Earth moved around the Sun instead of vice versa, but he still assumed that all motions were circular and at a constant rate. As a result, there still were differences between the positions he calculated and the actual ones, and he had to add epicycles to his calculations as well. After a sufficient number of epicycles had been added to bring the predicted and actual positions into tolerable agreement, his calculations were neither simpler nor inherently any more accurate than Ptolemy's.

This problem was resolved over the subsequent century. Galileo observed the phases of Venus for the first time through his telescope. Sometimes Venus appeared to be a crescent, and sometimes it appeared to be gibbous (that is, more than half illuminated as seen from the Earth). When a crescent, Venus was larger and—Galileo concluded—must have been roughly between the Earth and the Sun, so that only a small portion of the sunlit hemisphere was visible. In a gibbous phase Venus was smaller and thus must have been on the other side of the Sun, so that a larger fraction of the sunlit portion was visible. As Galileo well knew, we see crescent and gibbous phases of the Moon for the same reasons. These observations strongly suggested that Venus must orbit the Sun rather than the Earth. Galileo also saw for the first time the four large satellites of Jupiter that now bear his name as the Galilean satellites—Io, Europa, Ganymede, and Callisto. These objects moved back and forth relative to Jupiter and sometimes disappeared in Jupiter's shadow; they clearly were moving around Jupiter, which meant that the Earth was *not* at the center of all astronomical motions.

Kepler was unhappy with Copernicus's predictions of planetary positions because of the complexity of the necessary calculations and the unsatisfying requirement of epicycles. He set out to determine if a different type of motion would produce a simpler result. He assumed that the Sun was at the center of the planets' motion, as did Copernicus, but he tried orbits other than circular and velocities other than constant. He obtained an excellent agreement between the observed and predicted planetary positions when he assumed that the planets moved in elliptical paths in which the Sun was at one focus of the ellipse (with the other focus being empty) and when the speed of motion varied with distance from the Sun in such a way that an imaginary line between the planet and the Sun swept out equal areas in equal times.

The final piece of the puzzle came from Newton. His theory of gravity was based on each object being attracted to every other object, with the force of that attraction depending on their masses and the distance between them. He concluded that the gravitational force between two objects was proportional to the mass of each and inversely proportional to the square of the distance between them. By adding up the individual forces from every component of the Earth (inventing calculus along the way), he discovered that a natural consequence of this law acting on planetary bodies was orbital motions that exactly matched Kepler's assumed motions.

The combination of all of these pieces constructed a chain of logic indicating that the observed locations of the planets could be predicted just from the assumption that gravitational forces between planets controlled their motions in the sky. The approach embodied in this example is the canonical example of what has become the modern scientific method:

1. Locations and movements of the planets in the sky were predicted based on a simple conceptual model of how planets moved (initially, the Earth-centered circular motions and epicycles of Ptolemy).

2. The predictions were compared with a thousand years of direct observations of the planets' locations, and differences noted.

3. The disagreement was too large to accept, so the conceptual model was changed; the assumptions about the nature of planetary movement were modified—moving the Sun to the center, changing circular to elliptical movement, and changing constant motion to a specific law of velocity variations—and new calculations were made of where the planets would lie in the sky.

4. These new calculations were compared with the observations, and a much better match was found.

The end result was an improvement in astronomers' ability to predict the locations of the planets in the sky and an understanding of the underlying cause of the motions (that is, the role of gravity). More generally, an approach using basic physical principles to understand the relationship between cause and effect, by developing and testing hypotheses and then using them to predict events that have not yet been observed, became one major goal of science. As observations of a particular system improve or increase in number, further discrepancies can be noted and additional modifications of the hypothesis can be made. Over time, hypotheses about a particular system will improve and the agreement between predictions and observations will get better and better.

The final leap in the Copernican revolution required the acceptance not only of the concept of how the motions occurred but also that gravity was the actual driving force behind them. That is, we choose to equate the model that allows us to make predictions with a truth about how the universe actually operates. Motions of the planets were seen as the clockwork operations of a mechanistic system, and the universe became a system that was no longer incomprehensible; it could be understood based on simple principles. In other words, if we understand the basic operations of even complex systems well enough, we can predict from these "first principles" how the system will behave. An initial set of conditions describing a system, combined with an understanding of how the system operates, allows us to predict what the system will be like at any later time.

This approach has become the basic description of the scientific method. We make observations of a system, develop a hypothesis explaining what controls it, test predictions that follow from that hypothesis about how the system will behave by comparing the predictions with direct observations of the system, modify the hypothesis if necessary, and then start all over again. This is the empirical approach to science. The example of the Copernican revolution involves using a conceptual model of the forces that operate on a system to predict how the system will behave.

The same concept applies to laboratory experiments. An experiment can be set up in the laboratory in which an initial state of the system is carefully created, predictions are made as to what will happen when the experiment is run (that is, a prediction is made of the state of the system at a future time), the experiment is run, and the results are compared with the predictions. In this sense, experimental and predictive approaches describe the same concept of determining the relationship between cause and effect.

The Failure of Predictive Science in Astrobiology

The predictive approach may not work if a system is too complex to allow predictions to be made, is too large, or operates over too long a time scale to allow experiments to be performed or observed, or if random events that inherently cannot be predicted can have a large effect on the outcome. For example, we cannot use the "first principles" approach to understand how the planets in our solar system formed or how they have evolved in the 4.5 billion years since they formed. The planets are thought to have coalesced from a large number of smaller objects during the earliest epochs in the solar system. The ways in which these objects came together involved random collisions that cannot be predicted solely from an initial set of starting conditions. During the final assembly of the planets, seemingly trivial changes in the assumed conditions would have led to different outcomes (a given protoplanetary object might end up being accreted by the proto-Earth or the proto-Venus, for example), and therefore the final composition, size, and mass of the resulting planets would be impossible to predict. The final masses of the planets, from where in the protoplanetary disk the material that forms each planet came, and the history of impacts during the final assembly of the planets can be predicted only in a statistical or average sense.

Similarly, we cannot predict how a planet will evolve subsequent to its formation. The processes that control whether a given planet will have plate tectonics, for example, are not understood. How do they depend on the temperature distribution within the planet's interior, or on the composition and water content of the rocky interior? The history and geographical distribution of volcanic eruptions that will occur at the surface of a planet over the planet's history likewise cannot be predicted. They depend on too many factors that cannot be known and on random events that occur during the planet's history.

On the biological side, again we cannot use a first principles approach to understand the path of evolution during the history of life on Earth. The competition between individual organisms within a community or between communities of organisms is too complex to predict and depends on factors such as climate, weather, and, in the case of animals that have some level of sentience, individual choices that are inherently unpredictable. The fodder that drives evolution, in the form of mutations that occur in individual organisms, also cannot be predicted, although we do know that they can depend in part on the specific location at which a mutation-inducing cosmic ray is absorbed by an individual organism's DNA. The history of evolution also depends on nonbiological factors such as the random asteroid impacts that have plagued Earth throughout its history. Such an impact event killed off the dinosaurs (and a substantial portion of all living species on Earth) sixty-five million years ago; the resulting opening up of a large number of ecological niches allowed for the rapid evolution of a wide variety of different species (including those on the path that led to primates and, many tens of millions of years later, humans). What would have happened if that asteroid had landed entirely on land or entirely in the ocean instead of where it did, at the boundary between them? What if the asteroid had just barely missed the Earth, but a different asteroid had hit the Earth five million years earlier or later?

These are examples of events for which we cannot "do the experiment" by using a slightly different set of initial conditions and then observing the results or predicting from some set of initial conditions what the end result would look like. We cannot rerun the history of the Earth with a different mass, radius, water content, or distance from the Sun. Nor can we rerun the history of life using slightly different physical or environmental conditions on the Earth, with outcomes of competition between species being based on other random events, or with a randomly different history of giant impacts on the planet.

Certainly, we can speculate about what might happen under an alternative set of conditions. We can create computer programs, for example, that numerically simulate a simplified set of the processes involved in these complex systems and see how they behave. By running a wide variety of different experiments we can determine a possible range of outcomes. But these numerical experiments cannot include either the full complexity of the systems they are intended to model or the specific random events that must have taken place. While numerical models and computer programs can help us understand what might have happened, they do not tell us what actually happened in the specific case in which we are interested—be it the history of our solar system or the Earth, or the history of life on Earth.

Historical Science and Historical Narratives in Astrobiology

In dealing with the actual events that have already taken place in a system, we are looking at the past rather than making predictions about the future. We can look for evidence left behind by the past events and use it to infer what those events must have been. In deciphering the sequence of events from this evidence we are engaging in "historical science." That is, we use the inferred sequence of events to construct a "historical narrative" of what must have taken place in order to leave behind the evidence that we see today.

The impact event sixty-five million years ago that killed off the dinosaurs can serve as an example of how historical science works. Why do we think that such an impact occurred? It certainly was not an outcome predicted from a first principles approach to understanding the history of the Earth. There are objects orbiting the Sun that may hit the Earth at some time in the future, and there probably were similar objects in the past. And we know from the behavior of smaller objects (say, sand grain or pea sized) and from the occurrence of large impact craters on the Earth and Moon that impacts occur regularly on the Earth. The calculations that would allow us to predict specific impacts are inherently uncertain, however, and would be even if we knew the mass, composition, and orbit of every asteroid in the solar system. Rather, scientists think that an impact occurred then because they have found evidence left behind by that impact. There is a globally distributed layer of clay in the rock record of the Earth that has been reliably dated to have been deposited sixty-five million years ago. It contains an enhanced abundance of elements that are relatively depleted at the surface of the Earth and relatively abundant in asteroids and meteorites, such as iridium. The clay also contains abundant crystals of the high-pressure polymorphs of quartz known as coesite and stishovite; these occur in nature almost exclusively in debris that has been subject to the high-pressure and high-temperature environment of a hypervelocity impact. These measurements and facts are most consistent with the occurrence of a large impact. The most compelling evidence for such an impact emerged when scientists identified remnants of an impact crater that formed sixty-five million years ago and is big enough to have supplied the amount of iridium identified in the global clay layer. The combined evidence points very convincingly to the impact of a 10- or 15-kilometer-diameter asteroid with the Earth's surface sixty-five million years ago, producing a 180- to 300-kilometer-diameter impact crater and distributing debris from the impact globally. We don't need to have observed the impact to know that it happened.

Why do we think that the impact is responsible for the extinction of the dinosaurs? We see a dramatic change in organisms recorded in the fossil record at the time of the impact. Large numbers of species (including the dinosaurs) are found only in rocks that are older than this clay layer, and fossils of new species appear only in younger rocks. We have not found a fossilized dinosaur in the act of getting hit on the head by an asteroid, so we cannot be absolutely certain that this is what happened. However, we can make theoretical predictions as to what can happen in an impact of this magnitude: debris ejected by the impact and reentering the atmosphere would be heated to glowing hot temperatures that would trigger global fires; this debris would fill the atmosphere and block out all sunlight for perhaps several years; the heat from the impact might trigger the formation of acids in the atmosphere that would rain out in a heavy and destructive global acid rain. That is, we understand how processes that would be capable of leading to the extinction of a large number of species operate, and on that basis we can infer what happened. While this is not as compelling a case as the one we can build for the occurrence of the impact itself, and it still leaves the possibility that other events, possibly unrelated to the impact, actually killed the dinosaurs, the impact connections are strong and convincing.

We have used observations that we can make today of evidence that was left behind by events in the past to infer what must have happened to produce that evidence. Even though we could not have predicted this specific event in advance, we can understand the sequence of events that must have occurred in order to leave behind these geological traces. Geological events that have occurred throughout the history of the Earth are recorded in the rock record. When properly read, the rocks can tell us what the environment was at the time they formed as well as what organisms were alive. Rocks of almost all ages can be found, allowing us to read almost the entire history of the Earth from its formation 4.5 billion years ago up to the present.

We can take a second example of constructing a historical narrative from the biological sciences. We cannot predict from first principles the sequence of events that led to the origin of new species from mutations in existing species during the approximately 4 billion years since life originated on Earth. But we can look at the evidence left behind in the biological record and infer something about the sequence of the events that occurred.

All Earth organisms use DNA and RNA molecules to store and transfer genetic information. Sequences of nucleotide bases within each molecule contain the information necessary to build all of the proteins that catalyze chemical reactions within a cell; it is the sequence of these bases that really defines organisms and species. Changes in species occur through changes in the sequence of these bases, whether by incorporating new portions of a sequence, deleting existing portions of a sequence, or substituting one base for another as the result of copying errors or cosmic ray–induced mutations. As changes accumulate over many generations, in both the sequence of bases in the DNA and the functionality these represent, the organisms that carry them become different enough from the original organism to be classified as entirely different species. One parent species does not necessarily give rise to only one daughter species, however. Through multiple events that might include physical separation of populations of a species, several different daughter species can be created from a single parent species. In this sense, all life on Earth is thought to be derived from a common ancestor, after innumerable separations into different daughter species.

We can compare the sequence of nucleotide bases in two different species to see how closely related they are. If two species split off from their common parent very recently, their sequences will be nearly identical. The more time there has been in which to accumulate changes independent of each other, the greater the differences will be between the sequences. By comparing the sequences of a large number of species, we can determine which are closer together and which are farther apart. This information can be used to construct a "tree" showing the sequence of divergences of species from each other. It does not tell us the times at which they diverged, because not all species accumulate changes at the same rate. And it does not tell us the sequence of ancestors of a given modern species, because it looks only at those species that exist today. But it does tell us how closely species are related to one another.

Looking at the biological evidence contained within organisms that are alive today can tell us a great deal about the history of evolution as it actually played out on Earth. This is information that cannot be learned from a first principles understanding of how evolution operates, because the system is too complex. There is no way that we could predict the specific outcome of countless generations of evolution and separation into different species. The sequence of events (in this case, divergence of species from a common ancestor) is inferred from the evidence observed in living systems today. Our assembled inferences form a historical narrative of the events that must have taken place during the course of evolution. The fact that we did not "observe" the inferred evolutionary sequences as they occurred or perform the experiment ourselves does not affect our confidence that the historical narrative is accurate (within, of course, the uncertainties inherent in the technique).

Many of the questions addressed within the discipline of astrobiology and dealing with the potential and actual distribution of life in the universe cannot, in general, be addressed from a first principles approach. Nor can they be readily addressed through experimental or predictive science. It should be intuitively obvious that we cannot set up experiments to re-create the history of the evolution of species over the last four billion years on Earth, for example. Historical narratives based on the scientific evidence that exists today are the best way we have found to approach such problems. The field advances by continually searching for more (or more detailed) evidence—whether it be geological evidence of past events or of the history of fossils, or biological events recorded in the DNA and RNA of modern organisms—and applying logical processes to infer what sequence of events is most likely to have left behind these traces. A few examples will show how this concept is applied in astrobiology.

We cannot predict from first principles whether we will find life on Mars or Europa. While it is plausible that life is present there, it is also plausible that it is not. We will discover what is there only by going to Mars and Europa and searching. If we do find life on Mars, it may or may not represent an origin of life that was independent of the origin of life on Earth. I have already described processes by which microbes might have been transferred there from the Earth, carried inside rocks ejected from the surface of the Earth by asteroid impact. We cannot predict, in other words, whether any martian life we might find will be genetically related to terrestrial life. We can determine that only after the fact by examining its chemical characteristics: Does it use DNA and RNA molecules? Does it use the same suite of amino acids? Does it use ATP to store and transfer chemical energy?

What is the nature of planets around other stars? We did not expect to find Jupiter-like planets so close to their stars that their orbits take only a matter of days. We did not expect to find planets in highly eccentric (i.e., extremely elliptical) orbits so unlike the nearly circular orbits of planets in our own solar system. And we do not yet know the implications of these findings for rocky Earth-sized or Earth-like planets. Are such planets common or rare? Will they be absent from the systems that have the "hot Jupiters" orbiting close to their stars? Are they going to be common only around stars that do not have close-in gas giants, or might some as-yet-undetected process keep them from forming in those planetary systems?

If Earth-like planets are abundant, are any of them likely to have intelligent life? We have only our own single example of the development of intelligence, and it is hard to generalize from that. It is not clear that we will ever determine the events that were crucial in the development of intelligence here on Earth, let alone whether those events and processes are required in order for intelligence to form. Perhaps there are many pathways to intelligence. Some scientists today argue that intelligence is such an unlikely outcome, given the large number of random events that could have kept it from happening on Earth, that it is unlikely that intelligence will occur elsewhere. Others argue that intelligence is of such high survival value to a species that it will develop despite random events along the way. Clearly, we cannot predict from first principles how or whether intelligence will develop.

In each of these instances, we cannot predict the outcome using our basic understanding of what the relevant processes are and how they operate. The systems are so complex, and the outcomes so dependent on countless unknowable forces, that it will never be possible to predict them. We can, however, observe the properties and characteristics left behind by past events and infer what processes operated, the sequence in which they acted, and how the various processes interacted with each other. We can then use our observations to construct a historical narrative that describes the events. In many cases, the connection between the occurrence of a given event and the traces that are left behind is so strong that there can be little doubt about what happened.

In taking this historical narratives approach, one has to be careful not to give in to the temptation to create "just so" stories. An almost infinitely large number of scenarios might have created the evidence of past events that we observe today. However, we have the ability to search for additional evidence and to use what we learn to test the hypotheses we propose to explain events. It is this continual testing of hypotheses against the evidence, using logical and reasoned thought, modification or rejection of hypotheses, and a knowledgeable search for more evidence, that makes the historical approach a part of science.

Climate and the Potential for Life on Mars

Given the central role that Mars plays in astrobiology, we can use it to explore the use of historical narratives further. The standard wisdom describing Mars, as outlined in chapter 1, is as follows: Mars appears to have had stable liquid water at its surface in at least modest abundance early in its history, from about 4.2 billion years ago (when the surface geological record begins) until about 3.7 billion years ago. At that time the environment changed so that liquid water was no longer present or abundant at the surface. Liquid water has continued to exist within the martian crust, even in geologically recent times, and has been at least occasionally released to the surface. The long-term presence of liquid water, at the surface or in the crust, means that Mars meets the environmental conditions required to sustain an origin of life and/or its long-term existence. These characteristics make Mars exploration crucial to an understanding of planetary habitability and the distribution of life in our own solar system.

We can examine how these conclusions about Mars have been reached and the role that the historical approach to science has played in reaching them. We can also look at some of the issues that have been raised recently regarding the hypothesis based on these conclusions.

Let us examine the inferences derived from the morphology of the surface. Although there is evidence from other sources (such as the isotopic composition of the atmosphere), the appearance of the surface has been the primary driver behind our current understanding of the history of Mars.

The morphology of the surface is the result of past geological processes. Geological features that we see on Mars can be compared with similar features of known origin that occur on Earth. For example, the "valley networks" on the surface of Mars—branching systems of valleys that coalesce into larger valleys as they go downhill—have an appearance similar to terrestrial river drainage systems. We conclude by analogy that they formed when flowing liquid eroded the surface. The most likely liquid is water; other liquids are less abundant or less stable, and the valleys certainly look like they were carved by water (that is, their characteristics are very similar to valleys on Earth that were carved by water). Although liquid water is not stable at the martian surface today, these valleys appear to have been formed by the steady flow of water over long periods of time. This analysis suggests that the climate must have been warmer (and thereby allowed liquid water) when the valleys formed than it is today. As the valleys occur only on very old surfaces, indicated by the abundance of impact craters, we infer that the ancient climate on Mars must have been different from the present one.

A second morphological argument involves the erosion of ancient impact craters. Craters in the ancient (heavily cratered) portion of the surface with diameters larger than about fifteen kilometers have been severely degraded; they have lost the distinct blankets of material ejected by the impact that normally surround such craters, as well as the raised rims and central peaks that are common to all large younger impact craters and basins. Some of the most severely degraded craters appear as only a "cookie-cutter" imprint on the surface. The craters with diameters smaller than about fifteen kilometers are absent entirely. And some partially degraded craters have a large-scale "spur and gully" appearance in their interiors that is similar to features formed by erosion by liquid water on Earth. Again, these features suggest that they formed in an environment in which water was either more stable or more abundant than it is today. The ancient ages of these features based on impact crater abundance suggest a warmer and wetter environment on early Mars. In both cases, the features occur almost exclusively on surfaces that are thought to be older than about 3.7 billion years, and they do not occur at all on younger surfaces.

The simplest way for a climate to become warmer is by greenhouse warming of the atmosphere. Carbon dioxide, the major constituent of today's martian atmosphere, is an excellent greenhouse gas. Thus, the standard wisdom is that there was much more CO_2 in the martian atmosphere in early epochs, that it provided greenhouse warming, and that most of it was subsequently lost from the atmosphere. Trying to find where the CO_2 went—

for example, whether it was lost to space or formed carbonate minerals in the crust—has become an important part of the geological investigation of Mars.

Our inference that liquid water was present at the surface of Mars early in its history can be used to argue that life could have originated there. If life instead had been transplanted from the Earth inside rocks ejected by some enormous impact, it could have thrived there. Thus, these inferences, which are drawn almost entirely from the approach of historical science, have been important both in our understanding of the geological history of Mars and in driving our future plans to understand the habitability and biological potential of Mars. However, there are some problems with this scenario.

One issue is to explain how enough CO_2 could have been present in the martian atmosphere to raise temperatures enough to allow liquid water. Models suggest that the CO_2 would have saturated the atmosphere, so the atmosphere could not have been sufficiently dense to raise temperatures to the melting point of ice. It is possible that something else provided just enough additional warming to keep the CO_2 from condensing—dust in the atmosphere, perhaps, or the radiative effects of the clouds that would have formed. It also is possible that other greenhouse gases were present in sufficient abundance to raise the temperature without requiring such large amounts of CO_2. Either methane or ammonia would do, but these molecules would not be expected in any great abundance unless the life necessary to produce them was already widespread on Mars. It also is possible that globally averaged temperatures need not have been raised all the way to the melting point of ice. For example, liquid water could have existed underneath a colder ice cover, as can happen in Antarctica today, and perhaps still be able to form the water-related morphology.

A second problem is that at least one alternative mechanism for forming the erosional features has been suggested. The epoch of valley network formation and ancient impact basin degradation coincides closely with the early period of a higher rate of asteroid impacts onto the surface. Large impacts would have heated the surface, driving water into the atmosphere and heating the atmosphere enough (either directly or through subsequent short-lived greenhouse warming) to raise temperatures temporarily above the melting point. Water released from the subsurface would have rained out very quickly, but temperatures might have been so high that the rainfall was superheated to 1000°C or more, and any martian life would not be able to survive. This hypothesis also has some problems, involving, for example, the apparent contradiction between a predicted rapid origin of the drainage systems and the inferred slow origin, but these problems may not necessarily be fatal to the concept.

Thus, we have at least two competing hypotheses pertaining to the processes responsible for producing the present-day martian surface morphology. The two different hypotheses have very different implications for the early climate and the potential for life. A new piece of information comes from the Mars Exploration Rover *Opportunity*. Its landing site in the Meridiani Planum region of Mars was chosen on the basis of the presence of the mineral hematite in a coarse, crystalline form, detected using remote-sensing observations made from orbit. This mineral typically forms in the presence of low-temperature liquid water, and the mission was targeted there to determine whether liquid water had ever been present. *Opportunity* identified the source of the hematite as small, round concretions that we know form below the subsurface only when abundant liquid water is present. It also identified abundant magnesium sulfate salts that could have precipitated out when water evaporated from the surface or when water flowed through subsurface rock, and a type of cross-bedding in the layers of rock of a size and shape that strongly suggest that the sediments that formed it were deposited on the bottom of a stream or lake with flowing water.

Clearly, there is compelling evidence for liquid water in the Meridiani Planum region, both at the surface and in the subsurface. What is not known (yet clearly is relevant here) is the size of the body of water (lake? ocean?) or how long it lasted. Thus, although important, this information does not tell us which of the scenarios about the global climate is more accurate.

Is our understanding of the history of the martian climate about to undergo a paradigm shift? Is the standard paradigm facing problems so severe that it cannot be fixed and must be replaced by a different one? As we are still in the midst of these events, it is hard to tell whether we are in a Kuhnian crisis period. Many reasonable and well-respected investigators believe that the existing paradigm of an early warmer and wetter climate is fatally flawed and are actively searching for replacements. Others feel that the problems are not quite so severe, and that minor modifications of the existing paradigm will fix them. Multiple hypotheses have been suggested, and scientists are looking for arguments and evidence that support one above the others. Measurements that are pertinent to determining the temperature at ancient epochs are being defined, and they could be made from upcoming spacecraft missions.

Why Does It Matter What Type of Science Astrobiology Is?

It should be clear by now that many of the component disciplines of astrobiology use historical science. Many of the scientific issues astrobiologists deal with involve constructing historical narratives based on observational evidence rather than carrying out experiments or making predictions as to how a system will change over time. Most of our understanding of how the planets formed and how they function, how life functions and evolves, and the potential for life elsewhere are based on historical narratives. Many aspects of astrobiology are *not* historical, of course. The detection of planets around other stars and determination of their orbits and masses, understanding how iron- or manganese-oxidizing bacteria interact with their environment, and determining whether extant life is present in a given terrestrial (or martian) environment are all experimental or observational science. We need to recognize the distinction in order to understand how the different components of the discipline operate, how scientific conclusions are reached, and, perhaps most important for the scientists, how strong and certain those conclusions are.

The issue is not necessarily a straightforward one. In planetary science, for example, we build instruments that fly on spacecraft and make measurements of another planet, and we refer to this approach as conducting "experiments" there. Such studies often are not experiments in the classic sense of setting up initial conditions for a physical or chemical system and seeing what happens and then varying the conditions to see what effects that has. We combine the experimental and historical science approaches when we use these observations to determine the current state of a system and then use the current state to infer the past history.

Second, and closely related, the use of historical narratives in astrobiology is very different from the standard view of the scientific method that is taught at all levels in our schools, including the courses that professional astrobiologists teach: undergraduate nonmajors courses, upper-level courses for science majors, and graduate-level science courses. These courses teach the general public, future schoolteachers, and most future scientists about the formalism of science. To the extent that historical science does not actually follow the "scientific method" as it is taught, it is difficult to relate the theory to the practice. This discrepancy makes it hard for nonscientists to understand how science is done and what sets science apart from other intellectual endeavors; and it makes it hard for scientists to understand these distinctions in a way that allows them to discuss their approach to science with nonscientists.

An added complication here is that historical sciences can be forced into the mold of the scientific method. For example, again in planetary sciences, we can fly a spectrometer to a planet and make measurements from which we can infer the composition of the surface. From these measurements and inferences we can construct a hypothesis explaining what geological events occurred in the formation of the surface (that is, we construct a historical narrative). We can then use this hypothesis to make predictions about what else we might find on the surface, and then fly a new instrument that can determine whether the expected feature is present. Depending on whether it is or not, we can modify our hypothesis and start a new round of predictions. While this approach superficially fits into the mold of the scientific method, it clearly is distinct—in this case, we are choosing to measure different aspects of the present-day nature of the planet of interest rather than to carry out a different experiment and see what happens. Although these experiments expand our understanding of the current state of the system, and do so in a way that builds logically on our current understanding, this approach is fundamentally different from the experimental or predictive approach to science described above.

Finally, and perhaps most important, the use of historical narratives in science is important because it is through them rather than through experimental or predictive approaches that the "Big Questions" in science can be addressed: the ultimate fate of the universe; the origin of galaxies, stars, and planets; the history of life on Earth; whether there is life elsewhere in our own solar system or on planets orbiting other stars; the nature of intelligence and the psychology of the mind; whether we are the only sentient, intelligent beings in our universe. These questions are widely seen as the driving force behind much of space exploration, and they can be addressed *only* through the use of historical narratives. In none of these cases can we determine the outcome from first principles, laboratory experiments, or predictions based on our understanding of cause and effect. It is only through direct observation of the outcomes that have occurred, and inference as to what processes operated to produce them, that we can even begin to answer these questions.

5

Why Do We Do Science?

THE PRECEDING CHAPTERS DESCRIBE *HOW* WE DO science, as a way of understanding the nature of science and how the science of astrobiology works. In this chapter we consider the parallel question of *why* we do science. This question will lead to such fundamental issues as why the federal government funds scientific research, why basic research that has few if any practical applications (either in the short term or on the conceivable horizon) is supported, and why the public is so interested in questions about the nature and history of the universe. These questions will feed into issues of how to implement a research program in astrobiology. Our highest-level goals in exploring the solar system and beyond should come directly out of our understanding of the intellectual and programmatic drivers behind the program.

There is remarkably little discussion either among scientists or in classrooms about why we do science. What little discussion there is usually is informal and in the hallways rather than in formal classroom discussions. Such informal discussions provide few means by which the entire breadth of ideas can be discussed or the connections between science and the broader concerns of our society explored. Discussions at scientific conferences, such as those where NASA program managers describe the status of their programs, supplement these informal conversations, but there is little dialogue even there of how the individual programs fit into the broader scheme of the federal science program as a whole.

The question of why we do science immediately separates the various disciplines into those that address problems of practical relevance, such as finding a cure for disease, addressing global warming, and so on, and those that have little near-term practical relevance. As most of space science and astrobiology falls within the latter category, a few questions immediately arise: Why do astrobiologists engage in a research program that has few specific practical applications? How does the public view science, especially in terms of understanding it as a process rather than just as a static collection of facts? How does it view astrobiology, and what fuels the public interest in life elsewhere? What is the role of exploration in astrobiology? And, above all else, what would it mean to people on Earth to find evidence for life elsewhere? In this chapter we will examine the rationale for government support for astrobiology as it relates to public interest in the field and look at the reasons why the public is so engaged by the issue.

Science: The Endless Frontier

Scientists today tend to take the value of scientific research as obvious and the political environment of government support for research for granted. Top-tier universities and national laboratories are generally seen as the premier places at which to carry out scientific research. This is especially the case in astrobiology, which is driven by spaceflight missions carried out by NASA, by cutting-edge research that takes place primarily at universities, and by supporting efforts through NASA and the National Science Foundation (NSF). The NASA Astrobiology Institute is a good example. It was created in 1998 as a consortium of eleven (now sixteen) institutions working together to develop the discipline of astrobiology as a first-tier research initiative and to develop a cadre of scientists who could provide guidance to NASA on related flight missions and opportunities. Of the fifteen member institutions that were participating in early 2003, four were based at NASA centers (including two separate initiatives at NASA's Jet Propulsion Laboratory), nine were at research universities, one was at a national laboratory, and one was at a private research institution. Although each institution provided some support for the consortium's efforts, the bulk of the funding came from NASA.

Substantial federal funding for research has been available only since 1945, and few scientists in the community understand many of the details about the history or, especially, the rationale for the existence of this funding. Two generations of scientists later, few of the original participants are still active. To understand the basis for this continued support we need to understand the events that were taking place in 1945 and what has changed since then.

Toward the end of World War II, Vannevar Bush, the director of the Office of Scientific Research and Development, was asked by President Franklin D. Roosevelt to recommend ways in which the scientific achievements of the wartime era could be transferred to the public. The Office of Scientific Research and Development was responsible for technological development in support of the war effort, and under its auspices much of the technology that helped the Allies to win the war—from penicillin to radar to the atomic bomb (as controversial as this last is today)—was created, developed, and produced. Bush and Roosevelt recognized the necessity for science and technology development to continue after the war and were looking for ways to increase and enhance the government's investment in scientific research. The report that Bush created is a benchmark that set the national policy for government in science and science in government for years to come. Its title, *Science: The Endless Frontier*, refers to the role that continued expansion into a frontier had played in the development of the United States and the need for the exploration of the unknown through science to play the same role in the modern nation.

The basic conclusions of the report are striking in their simplicity: Continued progress in science and technology is of tremendous value to the country in terms of new products, new industries, new jobs, and so on. The benefits of such progress for public health, national security (as had occurred during the war), and the general public welfare are clear and, at least in hindsight, obvious. Contributions to this effort can come from research that takes place within government, industry, and academic institutions. Because each type of institution plays a different role in the development of science and technology, the government should support efforts within each venue. Finally, education of new scientists to participate in ongoing research and development activities is imperative lest there be insufficient numbers of well-trained persons to carry out the desired level of activity.

One of the report's most important recommendations was that the government should support a wide range of research activities. This range should include everything from basic research that has no apparent applications, and for which there is no obvious or predictable payback, to applied research that has immediate applications that quickly justify the investment of resources. Given its interest in short-term profit, industry tended to support research that offered short-term or immediate payback. Universities tended to support basic research that often involved allowing the researcher to wander in whatever direction was most intriguing. It was left to the government, then, to support basic research relevant to the missions of its individual agencies and research projects too large or too focused to receive funding from other organizations. And this support could be given to any of the types of organizations that do research—national laboratories, industry, or academia.

It may seem somewhat surprising that basic research was so highly valued at that time, given that it was the applied side of science and technology that had played such an important role during the war. Bush argued that while pure basic research might not have immediate applications, it would provide the fodder for applications down the road. There are numerous examples of basic research leading, decades later, to significant applications. The development of quantum mechanics and nuclear physics, for example, led to the atomic bomb. The discovery of the basic structure of the DNA molecule in 1953 led four and five decades later to gene therapy, better disease control through understanding of chemical interactions at the molecular level, and genetically altered, pest-resistant crops.

The Bush report had a tremendous impact that is still felt strongly today. The National Science Foundation was a direct outgrowth of the report, which offered detailed recommendations for the creation of an agency that would provide stable funding for basic research and education primarily at universities. The conclusions and recommendations listed above could have been written yesterday instead of six decades ago.

The report has guided the development of American science throughout the entire post–World War II era. The NSF and NASA have become mainstays of science support in areas related to astrobiology, to such an extent that science faculty members at major research universities are not promoted and given tenure unless they have substantial federal grants that support their research. Research funding that once came predominantly from private foundations or from universities themselves has been supplanted in many fields by funding from the federal government. In sum, Vannevar Bush's report, combined with the major role that science and technology development played in the outcome of the war, created an environment in which science and technology were highly valued and supported, often without question.

The conveyor belt of scientific research envisioned by the Bush report—from basic research to applied research to products and results that have a direct effect on our daily lives—was fundamental to his vision of the future of the country. Coming as it did just at the end of World War II, it is not surprising that the potential for military applications was given high priority in the plan. The Cold War added further impetus for federal support for science and technology. Military and political confrontation with the Soviet Union grew more intense in the years following World War II, and peaked with the building of the Berlin Wall in 1961.

Even the development of the space research program was arguably an offshoot of the Cold War. It was in essence a scientific and technological competition with the Soviet Union. In 1962, President John F. Kennedy started America on a race to the Moon with the Apollo program. In a speech at Rice University, he defined the nation's goals: "We choose to go to the Moon. We choose to go to the Moon in this decade and do the other things, not because they are easy, but because they are hard, because that goal will serve to organize and measure the best of our energies and skills, because that challenge is one that we are willing to accept, one we are unwilling to postpone, and one which we intend to win." Although Kennedy put it forward as a challenge to humanity and society, the program was clearly a challenge to the Soviets.

With the formal end of the Cold War, marked by demolition of the Berlin Wall in 1989, the role of science in the federal policy became unclear. Some felt that the primary justification for federal funding of research and the emphasis on basic research defined by the Bush report had disappeared. Several generations of scientists and citizens have come and gone since the publication of *Science: The Endless Frontier*, the national attitudes toward science and its role in society had changed dramatically, and the social issues were very different from what they were sixty years ago, so it seemed natural to expect that a new national policy on science might differ substantially from the earlier one.

Congressman Vernon Ehlers, co-chair of the House Committee on Science, led a study intended to update and redefine America's science goals in the post–Cold War era. The study's findings, titled *Toward a New National Science Policy*, were published in 1998. Perhaps surprisingly, the Ehlers committee reached many of the same conclusions as did Vannevar Bush. This similarity can be seen right from the beginning of the report in the committee's vision statement, which proclaims that "the United States of America must maintain and improve its pre-eminent position in science and technology in order to advance human understanding of the universe and all it contains, and to improve the lives, health, and freedom of all peoples." To achieve this vision, the Ehlers report recommends continued efforts in science as fundamental to the future of the United States and American society, and continued investment of national resources in order to be able to reap the benefits. In addition to the issues highlighted in the Bush report involving the military, health, and jobs, the new report also emphasizes the role of science in understanding, affecting, and protecting the environment and the role of an educated citizenry in being able to make policy decisions based on sound science.

One of the highest-level recommendations presumes that no other organization can or will take responsibility for basic research that has no immediate or obvious applications. It is up to the federal government to support this effort as an investment in America's future; and it should get the highest priority because it has no other source of funding. Most of the federal funding for science goes to support agencies with specific missions or objectives (such as NASA), which spend the money on their own projects. It is up to agencies such as the NSF to support broad-based fundamental research.

There is an interesting and important hole in the science policy, however, in both the 1945 Bush report and the 1998 Ehlers report. Neither makes any significant mention of the intellectual value of scientific exploration. The Bush report does not mention it at all beyond a statement of the importance of frontiers. The Ehlers report mentions exploration only in describing the vision that guided the committee ("in order to advance human understanding of the universe and all it contains") and does not expand on it elsewhere. Both reports emphasize the importance of basic research as a driver of the engine of progress, with applications to military, health, environmental, and economic issues. As we shall see, this failure to consider wonder, curiosity, and exploration is in striking contrast to the public's desire to explore the universe and search for and understand life within it.

Science as a Way of Understanding the World

Let us examine this other way of looking at science: as an intellectual endeavor and as a way of looking at the world. We need to begin with a discussion of how science is taught (as that determines how the public perceives science) and what the implications of that are.

The component disciplines within astrobiology—astronomy and astrophysics, planetary science, geology, and biology, to name a few—are often taught in universities to large classes of freshmen who are not science majors. This type of class is the most advanced formal science course that many college students take, and what they learn here about science is what they will take with them into the real world. Therefore, these classes represent an important (and, really, the last) opportunity to teach the educated public what science is, how it operates, and how it relates to the rest of their lives.

Typically, science at this level is taught as a static collection of facts. In geology, for example, it is easy to focus on the composition of rocks and minerals, the structure of the Earth's interior, the nature of volcanoes or of earthquakes, and the processes that go on inside the Earth and how they affect the surface. We usually teach these subjects as established facts, seldom even mentioning the measurements and observations that allowed us to establish the facts. In astronomy, we teach the facts associated with understanding the differences between the planets, the nature of stars and galaxies, and even the potential for life elsewhere. Professors find it easy to lecture on these topics, as they tend to be quite familiar with the basic facts and concepts in their own fields. And it is easy to test the students on the facts using true-false, multiple choice, or short essay exam questions that require little time to grade and little more than rote memorization on the students' part.

This "stamp collecting" approach to teaching science fails to recognize that the facts often tend to be the least interesting aspect of science, especially to students. The facts have a frustrating tendency to change with time, which should not be surprising given that one of the jobs of scientists is to discover new facts and place them into the context of the existing body of knowledge. And students have a tendency to forget the details, so that a couple of years later they will have retained very little of what they learned beyond a few "gee whiz" details. Perhaps most important, the facts have little intrinsic value in the absence of an understanding of where they came from and how they fit into an understanding of the broader aspects of the environment or the processes that control it.

A much more significant aspect of science, and one that is usually ignored completely in freshman-level courses, involves understanding the basic meaning and intent of science. One of the fundamental tenets of science, very seldom stated or discussed, is that the world is inherently understandable and that we understand how it works by observing it and by applying basic logic to what we observe. This point gets at understanding science as a process rather than just as results.

The work of Copernicus, Newton, and others in explaining the behavior of the planets in our solar system showed that the universe operates according to processes and forces that are understandable. The work by Darwin and his intellectual successors on evolution and the origin of species extended this concept to biology. Today, we understand that the characteristics of the natural world can be explained by natural processes and forces. Science is about finding out what these processes and forces are, and providing an explanation of how things work and how they have worked throughout time.

It is only through this approach that we can understand what distinguishes science from other ways of looking at the world. Furthermore, this "process" approach to understanding science allows us to get at the underlying relationship between cause and effect so that we can understand why the facts are what they are. It teaches us how to go from the observations that we make in the real world to the conclusions we can draw from them about how the world works.

Astronomy teachers have a truly outstanding example of science as the process of using observations to understand how the world works in the work of Copernicus, Galileo, Kepler, and Newton. Virtually every freshman-level astronomy textbook and every nonmajors course in the country discuss their names and contributions as part of the history of astronomy. But the significance of this particular example is not that it tells us how the discipline of astronomy evolved but rather that it represents a major step in the evolution of intellectual thought that has had tremendous consequences for the development of subsequent ideas.

The recognition that the Earth moves around the Sun was a triumph of observation, hypothesis, and explanation reigning over authority. For perhaps the first time, scientific observations and logic became more important than the edicts of the authorities. It was no longer necessary to take conclusions about the nature of the world on the word of others. Instead, people (either individually or collectively) understood that they could determine for themselves how the world works simply by making observations and applying basic logical concepts to interpreting them. In principle, anybody could become sufficiently well informed to understand the approach, the observations, and the interpretations pertinent to a specific problem and the strength of the conclusions and the uncertainties associated with them. This change in the way of understanding the world represented a fundamental shift in humankind's views of how the world works and marked the emergence of modern views of what science is. Today, four hundred years later, we still concern ourselves with the ramifications.

This point about the nature of science and the role of astronomy in the history of the development of science is not very widely appreciated or understood even by scientists. I have spoken with many professors who teach freshman-level courses or write the textbooks from which other people teach them. One colleague said that he teaches this idea implicitly by example rather than explicitly; the ideas of how science works are there for anybody who is willing to observe them. Anybody who has ever taught freshmen knows that one cannot teach anything implicitly, especially complex ideas that go beyond the memorization and recall of the basic facts. Another colleague told me that he would include these concepts in his book if I could tell him how to write exams to test students' understanding of them. This viewpoint ignores the idea that every textbook author or professor has the responsibility to decide what the fundamental concepts and ideas are and then to address them in a thorough and appropriate manner. If something is not in the textbooks and is not discussed explicitly in the classroom, then the professors have decided, whether consciously, by default, or out of ignorance, that it is not important.

Certainly it should be no surprise that undergraduate courses and textbooks do not discuss these issues adequately. Very few active scientists think about the broader questions of what science is or what the important concepts are; and as a result, they don't teach them. Their students eventually become college teachers responsible for teaching the next generation of teachers and scientists, and so each generation in succession continues to ignore some of the most crucial questions in science.

This idea of learning about the world by observing can be a subtle one as well. A different colleague commented that, in effect, we are all scientists. Every individual learns how to function in the world by figuring out how things work. We learn from experience to flip the light switch to the *on* position when it gets dark, to depress the clutch pedal before putting the car in gear, and that an airplane usually will get us there faster than a train. Although we may be applying experimentation to determine what is cause and what is effect (cause being the flipping of the light switch, and effect being the appearance of light, for example), this is not doing science. It provides us with no explanation of how or why the cause results in the effect, which in this context is the key feature of what science is. Teaching science requires that we teach our students to ask why things are the way they are.

The fundamental goal of science is to discover natural explanations for natural phenomena. We do not just characterize the behaviors and objects that can occur in the world; we also wish to understand the processes that control our world and the mechanisms by which they operate. In doing so we reaffirm that the world is inherently understandable and subject to successful inquiry.

Science as Exploration

With the above as a background, we can now approach the question of why society supports science in general and astrobiology in particular. I will start with a few basic observations about how science is done and the implications of that in informing our view of the nature of science.

My own research, for example, is aimed at understanding the history of the martian atmosphere and climate, the structure of the martian surface,

the nature of the geological processes responsible for producing that structure, the physics and chemistry of the martian surface, and the implications
of these issues for the potential for life to exist there. One of the notable points
about essentially all of these topics is that they have very few, if any, practical
applications; further, it seems unlikely that they ever will. I could say that
studying Mars will help us to understand the environmental requirements
necessary to support life or the range of environments in which life can exist,
and thereby help us to better understand the nature of life here on Earth. I
could also suggest that the life we might discover on Mars could help us to
find a cure for cancer or other diseases. These are weak arguments, however,
because the potential practical value of the results lies far in the future and
may never materialize. In any case, these are not the reasons why so many
people are interested in exploring Mars.

More broadly, few of the component disciplines within astrobiology have
any real potential for immediate practical relevance. Certainly there may
be some practical applications. By discovering all of the kilometer-sized and
larger asteroids in the solar system and mapping their orbits around the Sun,
for example, we could protect the Earth from future devastating impacts (as
seen in several really bad movies made during the last two decades). Understanding the behavior of the Sun and Sun-like stars may ultimately provide
knowledge about the history of the Earth's climate that could be used to
mitigate humans' influence on climates worldwide. The information gained
from studying the genome of terrestrial microbes, useful in inferring the history of life on Earth and the genetic relationships between organisms, also
adds to knowledge of the functions of DNA and RNA, how they operate, and
how they might relate to disease. While these and other aspects of astrobiology may have practical value, however, they too are not the driving force
behind astrobiology research programs.

A second point is that specific scientific results tend to become obsolete
very quickly, especially in astrobiology, as exemplified in the earlier discussion of Mars exploration. Each new mission is more sophisticated and takes
more detailed measurements than the one before it, and quite often the new
information completely upends the hypotheses the mission was designed
to investigate. By some estimates, the doubling time for new knowledge in
many science disciplines is much shorter than a decade. Thus, the lifetime
of an individual new result is measured in years, not in decades or centuries.

This rapid turnover in knowledge is not surprising, in that science works incrementally. Each new result contributes to the overall knowledge in a field, which then becomes the basis for future research. In some fast-moving fields, results are obsolete by the time they appear in print in scientific journals.

Certainly, some ideas have more staying power. The concept of evolution by descent with modification has survived for a century and a half, and even though the subtleties are still being debated it is unlikely that the basic concept will ever become fundamentally different from that described by Darwin. Similarly, the Copernican idea of the Earth orbiting the Sun is both so basic and so well understood that it is not likely ever to change. Very few ideas are at such a fundamental level, though; most ideas come and go.

A third point about scientific results is that we never seem to be satisfied with the answers we get to the questions we ask. It is a hallmark of science that each answer usually raises questions at a deeper level. We continue to send spacecraft to Mars, and to launch new and improved Earth-orbiting telescopes, because we can use them to address questions at a deeper level—questions that often arose as we were answering some of the earlier questions.

What do these observations tell us about the nature of science—or more specifically, the nature of astrobiology? Because so much knowledge becomes outdated so quickly, at least at the level of most practicing scientists, we must not be doing planetary science or astrobiology solely for the sake of the new knowledge that we obtain by answering specific questions. If we were, then most scientists would seem to be wasting their time; their specific results usually have little long-lasting value by themselves. Rather, the more significant aspect appears to be that we are doing it to engage in a process of continually learning new information about the world around us. Even though results may go out of date relatively quickly, taken collectively they influence the direction of thought in the field and help to determine the nature of new measurements, analyses, or spacecraft missions. That is, the scientific process, or the *search* for new knowledge, appears to be more important than the new knowledge itself.

The salient aspect of planetary exploration, in fact, seems to be the emphasis on the term *exploration*. In learning more about the Earth and the other planets in our solar system, about the other stars in the galaxy, their planets, and the rest of the universe, we are engaging in the act of exploring the physical universe. In learning more about the nature of life here on Earth and the potential for life to exist elsewhere, we are exploring the biological universe.

This concept of exploration is a major driver of astrobiology research. For example, we usually talk about the space *exploration* program rather than the space *science* program. Scientific results regularly appear on TV and in the newspapers. But it is the exciting *new* results rather than those that fill in the details that really grab people's attention; it is the new discoveries from *exploration* that feed that interest.

A couple of examples will substantiate this point. During the early days of human spaceflight in the 1960s, essentially everybody in the world followed the drama first of putting humans into orbit and then of sending them to the Moon. Who could not have been fully engaged by the countdowns of the early manned orbital spacecraft, by watching an astronaut be strapped into the tiny Mercury capsule to be sent off into space, not knowing what would happen or how the individual or the spacecraft would respond? Who was not moved by watching Neil Armstrong step out of the *Eagle* lunar lander and onto the surface of the Moon for the first time? Even the launch of the first space shuttle in 1981 held the attention of people around the world.

As space missions became routine, though, public interest and attention waned. By the time of the third Apollo mission to the surface of the Moon, public interest had dropped substantially. Today, the space shuttle and International Space Station (ISS) draw very little public interest; most people are not aware at any given time whether there is a shuttle in orbit or even that the ISS is occupied by humans. This is the natural outcome of the planners' original intent to make the missions routine and regular. Only when the human drama recurs does the public again pay attention. This happened in 1970 with the near loss of the *Apollo 13* mission to the Moon, and then again with the losses of the space shuttles *Challenger* in 1986 and *Columbia* in 2003. It is the human drama of exploration that gets people's attention rather than the scientific results the exploration produces. *Apollos 14, 15, 16,* and *17* did spectacular science, as evidenced by the tremendous quantity and quality of the data obtained from the returned samples. But by then the missions had become a case of "been there, done that."

The *Mars Pathfinder*'s exploration of the martian surface in 1996 is another example of the importance of exploration. *Pathfinder* was the third U.S. lander to set down on the martian surface, but it had something that the two Viking spacecraft in 1976 lacked: *Sojourner*, a small rover capable of moving about on the surface and making measurements at different locations. *Sojourner* was able to roll up to individual rocks and examine each one. The photos and movies it took held the public in rapt attention when they were transmitted back to Earth. And, perhaps of equal importance, the science and operations teams were very visible during the mission, and the public was able to see for perhaps the first time the tremendous individual and group achievements, excitements, and disappointments experienced by the researchers as they operated the rover and lander. The human (and mechanical) drama became a major component of the story. The same thing happened with the Mars Exploration Rover missions in 2004. Their Web sites got more hits during the first months of the mission than any other Internet event in history.

When President Kennedy set America on a path toward the Moon in 1962, he did not say that the intent was to get new scientific knowledge. Most people recognized even then that it was the competition with the Soviet Union to get to the Moon first and to demonstrate technological superiority that drove much of the space program in the 1960s and beyond. The public was not nearly as interested in this issue, however, as it was in the sheer drama of the exploration, the excitement of those first steps toward exploring the universe in person. Science wasn't a distant third place on the list behind politics and exploration; it wasn't even on the list.

Clearly, basic research in science is not only about exploration and discovery. It covers a broad range of ideas and topics—from exploration and discovery to characterization and categorization to explanation and understanding—and each of these is an important part of the process. But it is the exploration component that really catches people's attention.

The Role of Exploration

What is it about exploration that so engages the public? If we say that we are driven by curiosity, that is merely putting a different label on it; what is behind that curiosity?

Astrobiology and astrophysics address questions that are close to universal, both to scientists and to the public. We look for answers to questions about how the universe formed and evolved; how galaxies and stars form, evolve, and die; how planets form, behave, and evolve, and whether they are widespread; whether Earth-like planets exist elsewhere; how life originates and whether microbial life exists elsewhere; and whether intelligent life is unique, rare, or common in the universe. These questions touch us deeply as humans. They get at the basic issue of how *we*, both collectively and as individuals, relate to our surroundings.

Exploration is one of the hallmarks of Western civilization. Certainly, much exploration down through the ages has been driven by the prospect of economic gain or world conquest. The exploration of the New World five hundred years ago is a good example. Only when it became clear that there was a prospect of getting a good return on the investment was there a substantial expenditure of resources for exploring the Americas. Setting sail on a voyage around the world and going where nobody had gone before was not a way to get rich, though. Exploration had nothing to do with making a profit.

Our exploration of the universe has many of the same personal elements. We are interested in the earliest history of the universe because it explains how matter was created and evolved, and how our own constituents came to be. We are interested in the formation of galaxies and stars for the same reasons—we wish to understand our own star, which is just one of a large number of stars in the galaxy. We look for planets orbiting other stars because we are interested in the processes that were responsible for the origin of our own planet. And we are interested in the evolution of the other planets in our solar system as a way to understand the evolution of the Earth and why and how it ended up the way it did. Finally, we are interested in the origin and evolution of life on Earth, and the possibility that life might exist on Mars and elsewhere, as a way of understanding how we as humans came to exist, what our relationship is to the rest of the biological universe, and whether beings similar to us might exist on other planets.

In sum, we are interested in the things that touch most directly on our own existence and the things that describe the "boundary conditions" that surround our own existence. By learning about the universe we are learning about our own corner of that world. By learning about the nature of life

in general we are learning about the nature of life in a particular instance—
our own.

In this context, exploring the physical and biological characters of our
world is no different from exploring other aspects of the world around us. We
value the exploration of the arts and of the humanities as a way of understanding how we fit into our surroundings. We explore the nature of the "human condition" through literature and art. And we can explore the nature of
the human mind, of consciousness, as a way of understanding ourselves. In
effect, our explorations—as individuals, as societies, and as a civilization—
teach us about ourselves, about what it means to be human.

The Significance of Life Elsewhere

If it is the exploration that is important rather than its results, we can ask
what the significance would be of actually finding life elsewhere. Obviously,
the significance of finding life would have different meanings for different
people. Let me start from a personal perspective of what it would mean to
one individual scientist.

I believe that finding life elsewhere that had an origin independent of
the origin of life on Earth would have profound philosophical ramifications.
(The caveat that it be independent of terrestrial life is meant to exclude finding life on Mars that had been transferred there from the Earth inside a rock
ejected into space by an impact. While it would be intriguing to discover that
there was life on Mars that was genetically connected to terrestrial life, it
would not carry the same meaning as would an independent origin of life.)
Finding evidence for another origin of life, even the simplest or most primitive microbe imaginable, would tell us something fundamental about the
distribution of life in the universe. It would suggest that life is just another
form of the types of chemical reactions that can occur in a dirty planetary
environment (admittedly, though, a particularly interesting form of planetary chemistry).

A discovery of life elsewhere, in this view, would be at least as profound
as the discoveries by Copernicus and Darwin. The recognition by Copernicus that the Earth goes around the Sun had the effect of displacing the Earth
physically (and emotionally) from the center of the universe. Darwin discovered that the life distributed around the Earth today was not created in its

full-blown present-day diversity but is rather the end product of some four billion years of change and diversification, and that all life that exists here today is descended via modification and competitive natural selection from a common ancestor. The inclusion of humans in this scenario had the effect of displacing us from the superior position we had claimed for ourselves. Similarly, a discovery of an independent origin of life would displace all life on Earth from the center of the biological universe. Terrestrial life would become just one more example of the life that can exist in the universe. This would be the final piece of the puzzle necessary to tell us that there is nothing unique about humankind's place in the universe.

We can compare this view with the views of nonscientists. I teach a course on "extraterrestrial life" to undergraduates. The students in that class come from all across the campus (including the natural sciences, the social sciences, the humanities, business, and engineering). I asked one of my classes to tell me their views on the philosophical significance of finding life elsewhere. Typically, many felt that finding bacteria elsewhere would be scientifically interesting but that only finding extraterrestrial intelligence would have truly profound effects. Finding microbial life elsewhere, they thought, would not make any practical difference to most people; their lives would go on much as before. The philosophical implications might be substantial, but the students found it hard to imagine that there would be much change in people's day-to-day activities unless we were actually invaded by aliens. In this same sense, the Copernican revolution did not affect people's day-to-day activities; it matters little to an individual in any immediate practical sense if the Earth goes around the Sun or vice versa.

A few of the students felt that we should solve our own problems here on Earth before we go looking for life elsewhere. This comment gets at the underlying dichotomy between doing science for practical reasons versus doing it for its exploration value.

When asked about the ramifications of finding intelligent life, a number of students were confident that extraterrestrial intelligence will help us to save the world by solving all of our current problems. There are two interesting points here. Not only did some people believe that extraterrestrial intelligent beings would be benevolent and would work to help us, they also believed that our own problems are so great that we cannot solve them without outside intervention. One can think of this as the *Contact* perspective on alien life, harking back to the Carl Sagan novel and movie in which aliens interact with Earthlings in an attempt to guide us in the proper direction.

In contrast, many felt that contact with extraterrestrial life will destroy our civilization, either by intent or by accident—the *Independence Day* view of alien beings. This response reveals a real concern that extraterrestrial organisms might be malevolent and deliberately destructive; or that they might destroy us as an irrelevant by-product of some other activity, such as mining energy or resources from our planet. An advanced civilization with benevolent intentions might be just as harmful. Their science might be so far advanced beyond ours, and their art, philosophy, and intellect so sophisticated relative to ours, that they would completely subsume our own culture and thereby destroy it. The equivalent process has happened numerous times on the Earth, when previously isolated cultures at very different levels of technological sophistication have come into contact with each other.

These last two viewpoints together represent an interesting dichotomy. On the one hand, we might be saved by extraterrestrial visitors; on the other hand, we might be destroyed by them. In the absence of any real information on the psychology of extraterrestrials, one is free to read absolutely anything into the question. Rather than being a realistic assessment of the potential outcomes of interactions with alien life, it becomes an assessment of the psychology of the person holding the view, a litmus test as to whether one is an optimist or a pessimist!

There was a relatively common suggestion from the students that the existence of extraterrestrial life or intelligence would be inconsistent with the views of modern religions. Even a single discovery of extraterrestrial life that originated off the Earth would be inconsistent with the views promulgated by Western organized religions and would trigger their complete collapse.

The alternative viewpoint, also held by many, suggests just the opposite, that modern religions have adapted to deal with scientific discoveries and societal changes in the past and can probably adapt to deal with the discovery of life elsewhere (if there is even a conflict). Most theologians do not believe that there is any inherent conflict between the potential for life elsewhere and the basic belief systems inherent in religion (I discuss this issue in more detail in the following chapter).

One of the most interesting comments was that a discovery of life elsewhere would not be anything new. Some students suggested that we have already discovered extraterrestrial life and the government is hiding it, probably in a hangar near Roswell, New Mexico, or in Area 51 in Nevada. Although only a couple of students were willing to mention this viewpoint explicitly, other comments implied that this may be a widely held view. People are inclined to believe both that extraterrestrial intelligent life is visiting the Earth and that the U.S. government is willing and able to cover it up and keep it hidden.

The common thread here is that most people are interested in the issues surrounding the possible existence or absence of life elsewhere and are engaged by the ways in which such life might interact with humans. This is the case whether we are thinking about the potential for primitive microbes to exist or to survive on Mars, the potential for microbes or more complex beings to exist on planets orbiting other stars, or the possibility that aliens are visiting the Earth on a regular basis and, by the way, abducting people in order to conduct bizarre experiments on or impregnate them.

The students' interest in intelligent life but not in microbial life marks an important distinction. They valued the intellectual idea of multiple origins of life less than the idea of multiple origins of intelligence. The question of whether aliens would be benevolent or malevolent informed their interpretation, as did the view of how religion would be affected. Comments from several people suggested that the discovery of extraterrestrial intelligence is not as important as the subsequent interactions we would have with these beings. What would we learn from them, either scientifically about the physical world or socially about how to survive in it? Of equal significance, what would we learn about ourselves from the discovery of life elsewhere? As one student expressed her interest in contact with extraterrestrial life, with several layers of meaning, "We are looking for ourselves."

Implementing an Appropriate Space Exploration Program

We have considered two very different views of the role and value of science in our society today. In one view, it is the ultimate potential for practical applications that justifies efforts in basic research. In the other, it is the intellectual value that justifies it—the excitement of exploration with the ultimate value of understanding our world and ourselves. In this perspective, basic research, and especially research in astrobiology, is a part of "high culture" and would be an important part of any civilized society. The first view is the science policy of the U.S. government; the second is the perspective of many members of the public.

This dichotomy between science as engine of progress and science as intellectual excitement has been noted many times, of course. But it is not necessary to choose between these perspectives; they do not have to be in conflict. Rather, we can recognize science as contributing in both areas and as worthy of support for both reasons. It is possible to design a space sciences and astrobiology exploration program that selects specific targets in order to satisfy the need to explore and at the same time to design missions, instruments, and observations that do basic research of the highest quality.

Of course, not all space missions have to be aimed at finding life in the universe; nor do all of them have to seek splashy results. It is imperative, however, that every program be well balanced. Every mission should have a well-articulated overall theme or approach, and individual components should contribute to the theme in ways that can be understood and appreciated by the public.

The NASA Space Sciences program has changed significantly during the last five years in recognition of these issues. Astrobiology in the broadest sense has become the intellectual centerpiece of the NASA programs and encompasses a large portion of the space science activities. It is at the center of the 2003 NASA Space Sciences road map detailing research plans for the subsequent three years. The road map explicitly describes the connections between NASA programs and the broader issues of understanding habitability, the distribution of habitable planets throughout the galaxy, and the search for life off the Earth, emphasizing the importance of these questions in driving and justifying the entire Space Sciences program.

This strategy is in line with the emphasis placed on astrobiology in the decadal strategy for exploring the solar system published in 2002 by the National Academy of Sciences. That document clearly recognizes the importance first of determining the habitability of planets and satellites in our solar system and then of determining whether those places also contain evidence for life. These questions are now among the highest-level science goals of the solar system exploration program.

More recently, in January 2004, President George W. Bush announced a new vision for NASA that responds directly to the long-standing absence of a coherent rationale and plan for space exploration. The president began his announcement by noting: "This cause of exploration and discovery is not an option we choose; it is a desire written in the human heart. . . . Today, humanity has the potential to seek answers to the most fundamental questions posed about the existence of life beyond the Earth." He argued that the intellectual impetus behind the program would be the public's intense interest in finding out about life elsewhere. The "exploration" component of space science, he continued, specifically the drama and perspective of human involvement, would engage the public and "inspire us—and our youth—to greater achievements on Earth and in space." President Bush's vision calls for the expansion of human presence to the Moon and Mars and then throughout our solar system.

As is the case with many programs, though, the devil is in the details of the implementation. Implementing this new vision of human exploration of the solar system may not be possible within the available funding without cutting back the robotic missions that have been the mainstay of space science and astrobiological exploration for decades. And it is unclear to what extent Congress, the science community, and the public will support this vision, and therefore what kind of space program and science policy the United States will have for the next two decades.

6

Astrobiology, Science, and Religion

HOW WOULD THE DISCOVERY OF LIFE ELSEWHERE affect our religions and religious beliefs? What are the implications of such a discovery? This question really builds on a much broader list of questions about the relationship between science and religion in our society. Is there a conflict between science and religion? Are scientists antireligion? Is religion antiscience? To use the title of a recent book on this topic by philosopher of science Michael Ruse, can a Darwinian be a Christian? At a highly visible and pragmatic level, this discussion extends to the ongoing political issue of whether creationism (or its equivalents, "scientific creationism" and "intelligent design") should be taught alongside evolution in our public schools.

These questions arise because both science and religion are important in the lives of many people. Advances in technology and in our scientific understanding of the world have raised significant ethical and moral issues. At the same time, the vast majority of Americans profess to be religious. Most participate in some organized religion and try to live by the moral and ethical values of their religion.

Astrobiology has not yet had a major practical impact on religion, in large part because there has been no discovery of life off the Earth. A large fraction of the public believes, however, that a discovery of life elsewhere — particularly intelligent life — that is not genetically related to terrestrial life would have major, and devastating, implications for most modern religions. An obvious corollary would be that a search for life elsewhere that turned up nothing would have a significant positive impact on religion by suggesting that Earth life or intelligence is unique. Before we can address this question, we have to talk a little bit about the relationship between science and religion in general. Atheists and the nonreligious should also pay attention to these issues—the questions deal with the role of science in our society and with the interactions between scientists and society, and for that reason they cannot be ignored.

It makes no sense, of course, to talk about religion as a single, monolithic entity; there are many distinct religions. And a single faith can encompass a wide range of views. An individual may have views or beliefs different from those of the religion's leaders, which may themselves differ from the views of a theologian. Here, I will focus on the relationship between astrobiology and Western Christianity as a means to highlight the issues that can arise where science and religion intersect. My approach will be merely to identify and discuss some of the issues that arise that relate to religion and science rather than trying to resolve them in a definitive way.

Determinism in the Universe

Before considering the relationship between science and religion in general, we will examine the impact that several specific scientific discoveries have had on society over the last few hundred years. These examples should make it clear why major scientific discoveries have had significant implications for religious teachings and why the dividing line between them often has been described as a battlefield. We will focus on three scientific discoveries: the Copernican revolution, Darwin and evolution, and the development of quantum physics.

We have already discussed the Copernican revolution in several different contexts. The Earth goes around the Sun, and the simple physical processes embodied in Newton's law of gravitation explain most or all of the motions that are observed. With this discovery the motions of the planets, once describable only empirically, became orderly, understandable, and predictable. This change gave rise to the idea of a mechanistic universe that operated by now-understood rules. Given the clockwork-like precision of the system, accurate knowledge of the present state of it should allow scientists to determine both the system's entire past history and its future behavior. In this sense, the world was deterministic.

These ideas were thought to carry over to the realm of human existence as well. Humans could be seen as just another type of machine, subject to the same laws of physics that controlled the universe, with lives just as predetermined. How could humans exert free will if the processes governing the universe were unchangeable and the forces uniquely determined? Was free will real, or did we not yet understand the processes that controlled behavior well enough to allow valid predictions or projections? This viewpoint also gave rise to an apparent duality between the body and the mind or the soul—the former understandable in all its aspects, the latter unknowable and unpredictable. But the unknown might become known in time, and eventually humans might understand the full operations and extent of their existence. This mechanistic view seemed to leave little room for the existence of a God who could act in the present; the deity was instead consigned to the role of having created the universe and then allowed it to operate on its own.

The accompanying displacement of the Earth from the center of the universe also was a significant issue, and it raised questions about whether there were other "Earths" out there that might be occupied by other intelligent beings. Certainly, it called into question the special relationship that humanity had with God.

The second issue began in the 1800s with the publication of an idea that could explain the origin of species and the evolution of life on Earth. Prior to Charles Darwin's theory of descent with modification and natural selection, scientists had long recognized that the nature of living organisms had changed over time. The change could be seen in the geological record, with fossils embedded in rock layers differing from those in earlier and later strata. With the publication of *On the Origin of Species* in 1859, Darwin provided the now-accepted explanation for how species could change. He suggested that competition between members of a species, combined with the natural variation in any given characteristic within a species, would allow those organisms that were better adapted to their environment to reproduce more effectively and to pass on their traits to their offspring. Over long periods, organisms would accrue substantial numbers of changes and would become sufficiently different from their ancestors to be distinct species. By this process, new species could evolve from preexisting species. The species represented on Earth today are thus the end product of evolution over hundreds of millions of years. And humans, according to this concept, are just another animal species that evolved from preexisting, nonhuman species. In this context, the tremendous complexity and diversity that we see in organisms today are the result of relatively simple processes operating over long periods.

In contrast with the mechanistic and deterministic world described by Newton, Darwin's history of biology could be seen as having unpredictable twists and turns. While not strictly a random process, evolution is an emergent process, and life an emergent system that cannot be predicted even from a full and complete understanding of the system's components. It was not obvious what role God could, would, or should play in this seemingly random and somewhat arbitrary universe. If species originated through natural events such as mutation and natural selection, then the origin of humans would seem to have been the result of a random series of events that had little connection to God. The Earth could not have been created in six days less than ten thousand years ago. Application of the concept of evolution to the descent of humankind created a firestorm of controversy over the intellectual and pragmatic relationship between the science of Darwin and the religion of the Bible (as manifested by the Book of Genesis, for example) that continues today.

Finally, the early part of the twentieth century saw the development of what has been termed modern physics—and in particular of quantum physics. The atom was now understood to absorb and release energy only in discrete amounts rather than in arbitrary amounts. One of the implications of this theory was the understanding that light can behave simultaneously as both a wave and a particle. Only if a photon of light has the right energy levels, for example, can it eject electrons from the surface of a metal, independent of the intensity of the light beam (i.e., the photoelectric effect, for which Albert Einstein received the Nobel Prize). Not only can light be both a wave and a particle, but things that we normally think of as particles, such as electrons, also can exist in both forms. This duality introduced an inherent uncertainty into their properties. We cannot determine, for example, both the location and the momentum of an electron with infinite precision. If we know the location exactly, we cannot know the momentum at all; and if we know the momentum well, the location is uncertain. It is not a matter of our inability to measure things but rather of the inherent inability to determine uniquely something governed by statistical probability.

These results changed our mechanistic-deterministic view of the universe. Because of the indeterminate nature of processes at the subatomic level, outcomes can be known only with certain probabilities rather than precisely. And if the behavior of electrons and atoms cannot be predicted, then neither can the behavior of the larger entities that comprise them. This result had interesting implications for the human mind. On the one hand, it opened up the door again for the existence of free will, in the sense that all future behavior was not predictable from the current state. At the same time, it left open the possibility that human consciousness still could be an emergent property resulting entirely from the inanimate physical characteristics of matter. And it also left room for God to act in the present-day world through behavior and events not fixed in advance.

With this background, we can now ask about the different ways in which science and religion can interact, and how we might be able to reconcile our current understanding of the physical and biological world with our views of religion. And we can ask what the potential interactions are between astrobiology and religion. Again, these issues are important for the nonreligious as well, as they relate to the broader societal desire to understand the connections.

How Does Science Relate to Religion?

The relationship between science and religion is often seen as a battle, with the very future of humanity and its moral and ethical development at stake. Like so many things in modern society, this viewpoint goes back to the universe described by Copernicus and Newton. Although Copernicus was the first of the modern astronomers to suggest that the Earth orbits the Sun, his book was not published until he lay on his deathbed. Galileo, on the other hand, pushed the idea of a Sun-centered universe very publicly (and in Italian, which was much more widely read than the traditional Latin), argued his case politically, and ended up imprisoned for heresy for his efforts. The Church did not officially exonerate him until some four hundred years later! The concept of a battle between science and religion was exacerbated in the late nineteenth century with the rise of the idea of Darwinian evolution. At the turn of the twentieth century, a number of widely read and influential books described the "war" between science and religion and the new science that appeared to rule out any significant role for God.

This view of "science versus the Church" continues today among the public, even if not among theologians. It is exemplified by the very divisive current debates about the teaching of evolution and creationism in the public schools. "Evolutionists" argue that intelligent design is not science and should not be taught in science classes. "Creationists" argue that a literal interpretation of the Bible (and of Genesis in particular) is correct and that their view of the history of the world and of terrestrial life is at least on an equal intellectual footing with the evolutionary view and should not be discriminated against. Individual school boards around the country are being asked to make decisions about what is and is not to be included in the public school curriculum. In this context, it really is a battle. If the supporters of evolution win, Christian Fundamentalists will lose the opportunity to have the origin of humankind as described in Genesis given the same level of credibility as the scientific view, and intelligent design creationists will have lost the opportunity to teach that evolution is guided by the hand of God. If the supporters of Christian Fundamentalism win, scientists will lose their ability to determine what science is and which scientific theories deserve acceptance. Interestingly, roughly half of the adults in the United States accept one of the creationist views of the origin of humans, and roughly half of the remaining adults do not fully accept that evolution occurs as the mindless, seemingly random process described by biologists; thus, only a relatively small fraction of the public accepts the history of life on Earth as put forward by modern biologists.

Is it possible for science and religion to coexist? How can the two interact and survive intact? One way is for there not to be a relationship. Science can be seen as a way of understanding and describing the operation of the physical world, and religion can be seen as applying to the world's spiritual, moral, and ethical aspects. In this way, there is no overlap between the spheres governed by each; each describes different issues and concerns, and as a result there need be no inherent conflict between them. This is the approach taken by evolutionary paleontologist Stephen Jay Gould, for example, in his 1999 book *Rocks of Ages*. It also is the view espoused hundreds of years ago by Descartes, who considered the human body a machine and consciousness a separate entity. Even farther back, around AD 400, Saint Augustine of Hippo, one of the great early theologians, hinted at the same idea. He noted the contrast between the "City of Man" and the "City of God" as a restatement of Christ's warning to Peter to remember the difference between what belonged to Caesar and what belonged to God. Saint Augustine argued that Christianity in particular lay within the heart and soul of the individual and therefore was not dependent in any way on the world of Man. In the face of a material world in which life was becoming increasingly difficult at the time of the decline of the Roman Empire, this distinction allowed followers to preserve the concept of the spiritual world. Saint Augustine concluded that Christianity, to survive, had to renounce earthly glory and be willing to live on in the individual; the fall of Rome, therefore, would have been of little consequence, as the goal of Christians would be met in a different life.

This idea of separation of the two realms is widely accepted today. Natural explanations are proffered for many observable phenomena. And divine intervention is often invoked to provide an explanation only where science has not provided one. However, this approach can give rise to its own problems. The world of science has a habit of continually encroaching onto the world of the spiritual. At any moment, our understanding of the world is necessarily limited and incomplete; behavior that is not explicable by science is attributed to divine intervention. As science progresses and new understandings are obtained, some phenomena once assigned to the realm of God are reassigned to the realm of science. God has increasingly become a "God of the gaps," and the gaps not explained by science are getting smaller and smaller. This process naturally creates tension between science and religion, and fuels the idea that they are in constant battle.

Saint Thomas Aquinas (d. 1274), the leading theologian of the middle ages, offered a different perspective. He tried to unite the City of Man and the City of God, arguing that there was only a single reality and that the physical and spiritual realms were merely different aspects of that one entity. As there was no way to separate the two in a clean, concise, and intellectually defensible manner, and no reason to separate them, they were essentially indistinguishable.

This view of unification was expressed again most recently by Pope John Paul II in his 1998 encyclical *Fides et Ratio* (*Faith and Reason*). The title of this missive reflects the fundamental issue it addresses: what is the relationship between those things known by faith and revealed truth and those known by applying reason to understanding our physical world? It argues that the two concepts are different ways of looking at the same reality and that they can never be in true conflict because there is only a single reality. The Pope quoted the First Vatican Ecumenical Council in stating, "Even if faith is superior to reason there can never be a true divergence between faith and reason, since the same God who reveals the mysteries and bestows the gift of faith has also placed in the human spirit the light of reason. This God could not deny himself, nor could the truth ever contradict the truth."

He went further, though, in insisting that both faith and reason are necessary components of any deep understanding of our world:

Deprived of what Revelation offers, reason has taken side-tracks which expose it to the danger of losing sight of its final goal. Deprived of reason, faith has stressed feeling and experience, and so run the risk of no longer being a universal proposition. It is an illusion to think that faith, tied to weak reasoning, might be more penetrating; on the contrary, faith then runs the grave risk of withering into myth or superstition. By the same token, reason which is unrelated to an adult faith is not prompted to turn its gaze to the newness and radicality [*sic*] of being. This is why I make this strong and insistent appeal—not, I trust, untimely—that faith and philosophy recover the profound unity which allows them to stand in harmony with their nature without compromising their mutual autonomy. The *parrhesia* of faith must be matched by the boldness of reason.

This view also builds on the ideas of John Calvin, who argued that Scripture cannot be taken as alone representing the truth. If nothing else, the existence of two very different versions of Genesis indicates that neither should be read as strictly factually correct. Calvin argued that revealed truth from thousands of years ago had to be considered in the context of what was understood about the world at the time, and so could not accurately or fully describe the physical world in a way exactly consistent with present-day scientific understanding. In other words, interpretation of Scripture has to evolve as our scientific understanding of the world around us evolves. In this approach, the ideas of the Church are based on contributions that come equally from Scripture, from interpretation of Scripture through the reasoning of humans, and from observations and analyses of the physical world.

No one approach should be taken in isolation; each contributes to a greater understanding of the whole.

Is Science a Religion?

Some argue that science itself is a religion—that it requires a faith in the scientific method as a way of understanding the world, and that its proponents have a faithlike belief in the validity of a scientific description of the history of the world that is comparable to belief in a religious description. Is this concept valid?

The usual answer from the scientific community is that this viewpoint is flawed. Christianity is based on revealed truth, which can never be questioned or tested independently. And its core beliefs are unchangeable, even in the face of uncontestable observations of the real world that might be in conflict with them. Religion requires a faith which by its inherent nature is a belief that is not based on objective observations. Science, on the other hand, is subject to continual checking by independent analyses. It is based on observations made in nature or in the laboratory and on hypotheses that can be discarded if they are found to be in conflict with the facts thus obtained. Science does not require a "revealed truth"—that is, taking somebody's word about anything—but can be carried out independently by anyone who is trained in the scientific method.

It is important to recognize that this standard response is a cartoon caricature of the ways in which we characterize both science and religion and how they function as intellectual and emotional endeavors. Certainly, religion is based on faith, and the core aspects of that faith are unchanging. However, the complementary nature of faith, with its reasoned interpretation, allows key aspects of religion to change with time in a way that accommodates our continually changing scientific view of the world, as discussed by Calvin and described above. This flexibility not only serves to keep religion dynamic but also allows it to stay in touch with modern times.

And, of course, science is not purely objective and based solely on reason; the human factor plays an important role in the development of scientific thought. There are numerous examples of scientists who espoused a particular explanation for something because they *wanted* that explanation to be true, and did so in the face of contradictory or problematic observations. Furthermore, science is not without its own dogma—statements that are widely if not universally accepted as true but have no underlying justification. These might include the physics concepts of conservation of energy (or of mass-energy in the modern system) or conservation of momentum.

At its most basic level, though, science is based on the *assumption* that the world is inherently understandable and that there is a natural explanation of cause and effect for everything that we can observe. Scientists assume that if we cannot identify a natural explanation for a particular phenomenon that is only because we have not found it yet. This view of a materialistic world goes back to Newton and Darwin and is deeply embedded in the implicit characteristics of science.

Investigating the origin of life on Earth is one example of this concept. Chemists and molecular biologists are trying to understand the processes that led to the origin of life by studying individual chemical reactions that can lead to increasingly complex molecules capable of carrying out certain catalytic functions. Specific molecules are seen as central to the earliest history of life, based, for example, on their deeply embedded nature in biochemical reactions and cycles in all domains of life. Scientists in laboratories study chains of reactions that might lead to these molecules being produced naturally in a prebiotic Earth, and mimic the formation in geological environments of RNA-like molecules that might lead to an "RNA world." These analyses are based on an implicit assumption that there exists a series of natural steps that would lead from a purely geological-geochemical environment that had no life to a biological environment. Our failure to find an appropriate suite of chemical pathways to date is not considered to be a problem but merely an observational statement that not enough research has been carried out yet. In this sense, the modern view of biochemical reactions leading to life does not go back very far—only to the ideas of J. B. S. Haldane and A. I. Oparin in the 1920s and 1930s and to the experiments of Stanley Miller beginning in the 1950s—and there just has not been sufficient time to explore enough potential biochemical pathways to have found plausible and appropriate ones.

Progress in the field of astrobiology relies heavily on uniformitarianism, the concept that the scientific laws we observe or infer to apply locally today also apply at other locations and at other times. In geology or planetary science, this concept implies that the processes that are operating now on a planet are the same ones that have operated at other times, although possibly at different relative or absolute rates, and that the features that we see reflected in the geological record are the result of the same forces and processes that we can see operating today. Only when known processes are demonstrated to be inadequate are scientists generally willing to invoke new or unknown processes. This is also the concept underlying William of Ockham's canon ("Ockham's razor") that the simplest explanation is to be preferred over excessively complex ones. This is the justification underpinning essentially all of planetary geology, and it allows us to observe morphological features on other planets and infer what processes acted to produce them (if it looks like a volcano, it probably is a volcano). But there is no fundamental reason why these concepts or laws have to be true, other than the fact that this assumption appears to work.

There are exceptions. The laws do not apply, for example, in the very high-temperature, high-pressure environment that existed just fractions of a second following the singularity at the time of the Big Bang. The exceptions are generally ascribed to their being in a physical regime different from the one we see at present rather than to a failure of the general laws per se.

Is science a religion? No. Despite its adherence to rules that are accepted on faith, it really does fall into a different class from religion. Although religion cannot be easily defined, most people understand that it deals with moral and ethical views, involves beliefs that are held for personal or individual reasons, and often connects these views and beliefs to the existence of a supernatural being or entity that governs the world. While there are some areas of overlap, science generally does not deal with moral and ethical issues or tell us how to resolve them. As many have described the difference, science deals with the "how" questions and religion with the "why" questions. Or, as one scientist put it, astronomy tells us about the heavens while religion tells us about getting to heaven.

Religion and Astrobiology

Now we get to the real crux of the discussion: What would the implications for religion be of a discovery of life elsewhere? This is a complex issue that, like many areas of religion and philosophy, has not been fully explored, and once again I will emphasize some of the questions here rather than attempting to supply detailed answers.

Do We Live in an "Anthropic" Universe?

A number of people have pointed out that many of the physical parameters that govern our universe appear to have been very carefully tuned to allow life to exist. These parameters control the ways in which components of atoms interact with each other. They determined the original amount of hydrogen that formed as the universe cooled down in its earliest history following the Big Bang (and whether any hydrogen would even be formed at all); the ability of hydrogen atoms to fuse together to form carbon atoms in the interiors of stars at a rate fast enough for carbon to form but not so fast that all of the hydrogen would turn quickly to carbon, leaving no fuel to power the stars; and the ability of gravity to allow matter to accumulate into structures such as galaxies, stars, and planets. Had values for the strong coupling constant, the weak coupling constant, the electromagnetic fine structure constant, or the gravitational fine structure constant been even a tiny amount different from their actual values, the evolution of our universe would have been fundamentally different and life as we know it would not have been possible. For instance, a slight change in the values would have left an entire universe filled with nothing but hydrogen unable to chemically combine to form any heavier elements. Life could not be constructed in a hydrogen-only universe. Similarly, if the gravitational constant had been only slightly different, gravity would have slowed down the expansion of the universe very quickly, and all of the matter would long ago have collapsed back together again into a singularity; the universe would not have existed long enough for life to form. Yet, there is no apparent fundamental physical law or reason why these parameters had to have the values that they have. It thus appears that our universe is constructed in an incredibly improbable way (that is, with very specific and fine-tuned values of physical constants), and in the only way that allows the existence of life. Was the universe created in anticipation of an origin of life? Does it require a thoughtful creator?

One possible response to this question involves the potential for multiple parallel universes ("multiverses") that comes out of superstring theory. The ideas this theory generated postulate an infinite number of parallel universes operating in up to a dozen different dimensions (of which we can see only three); each universe is incapable of interacting in any way with any other, and each presumably could have its own physical laws and numerical constants. While all combinations of constants might exist in one universe or another, life could form only in the ones in which they happened to have just the right values. With an infinite number of universes as a starting point, there would be an infinite number of habitable universes even if only a small portion of parameter combinations could result in life. Thus, for us to exist and be able to ask this question, we would of necessity have to be in a properly tuned universe.

A closely related question is why is there a universe at all? Why is there something instead of nothing? This question cannot, of course, be investigated by studying the known universe and would therefore appear to be outside the realm of science entirely.

*Are the Origin and Evolution of Life Consistent
with Christian Theology?*

The scientific view of the origin of life is that it resulted from natural chemical processes operating over hundreds of millions of years on the early Earth, some four billion years ago. This view is not consistent with the strict and literal Fundamentalist interpretation of Genesis, which claims that the Earth and everything else in the universe was created within the last ten thousand years, and every species of life on Earth resulted from an independent creation event. No amount of arm waving or attempts at reconciliation can bring these two views into alignment.

Is the scientific view consistent with the "accommodation" view of Christianity, which considers the Bible to be filled with metaphor that must be interpreted in the light of our modern understanding of science? There seems to be no fundamental inconsistency here. Saint Augustine, for example, put forward an interpretation of Genesis that had life arising out of the preexisting building blocks of the earth, the oceans, and so on. That is, the characteristics of life must have been built into the preexisting matter, and life could have arisen in due course on its own. On the face of it, this view does not seem so different from the scientific view of the chemical origin of life, in the sense that the latter sees the chemical reactions and the molecules that they can produce as resulting from the properties of the atoms themselves. Saint Augustine may have had something very different from evolution in mind, but the basic idea that the characteristics of life were built into the matter and that no further divine intervention was required in order for life to arise appears similar. The words are different, of course, reflecting the changes in our understanding of chemistry and biology over the last sixteen hundred years.

If evolution is understood as the result of purely random mutation and adaptation events operating over long periods, then it does not appear to be consistent with the Christian view of the history of humankind. How could we have been created in God's image, and how could humans be at the center of God's creation, if we are the end result of a string of accidents that occurred over a period of four billion years? If evolution acts on random mutations, how can God have intended to create us? And how can God intervene in our lives and on our behalf if he merely created the universe and then left it to run on its own?

One possible answer to these questions involves God acting through the openings allowed by the indeterminate nature of quantum physics. The Heisenberg uncertainty principle refers to the fact that the outcome of events at the subatomic level cannot be predicted except on a statistical basis. Thus, God has the option of operating within the constraints allowable by quantum physics. As mutations occur one by one at the atomic level in DNA and RNA molecules, statistical events can produce various outcomes. God can intervene to produce specific mutations. Thus, while individual mutations might seem to be produced by random effects, such intervention over time would have resulted in the evolutionary history that we see in the geological and biological records. The amount of time involved in the evolutionary sequence leading to modern humans is not seen as problematic, because God operates outside of time; in essence, four billion years is the blink of an eye to God.

While Fundamentalists object strenuously even to the idea that evolution took place, the official Roman Catholic view offered by the Pope is that the evidence for evolution by natural selection is overwhelming and cannot reasonably be questioned. The distinction that the Church makes is that faith requires that one accept that at some point God infused man with a soul.

Would the Discovery of Life Elsewhere Be Consistent
with Christian Theology?

We begin to get into more interesting, but of necessity more speculative, questions at this point. If the Augustinian interpretation of Genesis is correct, then those same characteristics that allowed life to originate on its own under the proper conditions on Earth were presumably built into preexisting matter that formed planets around other stars. Life would be just as likely to originate there under the appropriate environmental conditions.

According to the Fundamentalist view, however, if life exists elsewhere, then it too was created species by species. The Fundamentalist view raises other questions: If everything on Earth was created for the benefit of humans, what do we make of planets and life created off the Earth? Is the potential for our eventual discovery of them itself of sufficient value to humankind to justify their existence in a human-centered universe? If God "merely" bridged an otherwise unbridgeable gap in the chemical origin of life on Earth, is it likely that he did the same thing innumerable times on innumerable worlds throughout the universe? The more widespread we eventually find life to be throughout the universe, the less plausible this would seem to be.

Can Extraterrestrial Intelligence and Christian Theology Coexist?

Two of the underlying issues here involve the relationship of the extraterrestrials with humans and the existence or necessity of a Redeemer in the form of Jesus Christ. One aspect of Christian thought suggests that the Earth and everything on it exist for our benefit, as do the heavens. If that is true, then stars would of necessity be populated with planets, and planets with life and intelligence. This view is embodied in a comment in the 1997 movie *Contact*, about a first encounter with intelligent life beyond the Earth. Dr. Ellie Arroway (played by Jody Foster), a radio astronomer involved in the search for extraterrestrial intelligence, suggests: "If there isn't other intelligent life out there, it would be an awful waste of space." Again, what is the value to humans of the existence of intelligent life elsewhere?

An alternative perspective on Genesis is that humans were created as stewards of the Earth. In this context, extraterrestrials could be seen as stewards of their home planets, and there is no inconsistency. Issues arise only when one or more civilizations eventually venture out and come into contact with the others; the nature of any problems depends, of course, on what actions are taken on contact.

The question of the Redeemer is more complicated. Would intelligent life elsewhere have free will and necessarily undergo the same fall from grace that Adam and Eve did? Does the human redemption that accompanies the death of Jesus Christ on the cross redeem all life throughout the universe? Would Jesus's redemption here on Earth carry outward as humans explore the universe, spreading Christianity throughout the galaxy as they come into contact with intelligence on other worlds? Can extraterrestrials be redeemed through us? If Jesus's death does not redeem the entire universe, is it conceivable that God would require a Savior to die on the cross on each world, and is this physically and metaphysically possible? Could different mechanisms of redemption occur elsewhere? Christianity has already dealt with the issue of spreading Christianity on Earth, in the form of people who existed prior to the time of Jesus or were from non-Christian cultures (such as Polynesians) that were introduced to the form of Jesus much later.

All such questions and discussions have considerable uncertainty. As expressed by Ernan McMullin, a philosopher and theologian from Notre Dame University who has dealt with many of these issues, "If our imaginations can scarcely encompass such features of our cosmic home as action at the quantum level or the first moment of cosmic expansion, we should be modest in what we have to say about the Creator who set those limits in the first place."

Science and Religion Revisited

Is it possible for a scientist to be religious? Does accepting science require rejection of religion? It might seem so at times. Scientists, especially some of those working in the areas of the origin, evolution, and fate of the universe, are prone to making pronouncements about the implications of their scientific results for religion. Physicists have written books stating that the physical laws of the universe are the only agents operating. We have always found natural explanations for natural phenomena, they say, so no other explanations can exist and there is no need for the existence of God. In this same vein, cosmologists have written books about "the God particle" or finding the "face of God," arguing that only through the physical laws of the universe can God possibly exist and operate.

Surveys of scientists show that they tend to be nonbelievers at a much greater rate than the general population. There are believers among scientists, though, so there must be ways of reconciling the two worldviews. Kenneth Miller, a microbiologist from Brown University who has been very active in promoting the scientific validity of evolution in the ongoing debate about teaching intelligent design in the public schools, has discussed one way to do this. In *Finding Darwin's God* (1999) he points out that intelligent design and creationism are fundamentally at odds with a scientific understanding of the world. He further argues that one cannot pick and choose which canons of science to accept and which to reject, and that a rejection of evolution is, in essence, a rejection of all science. A complete rejection of science is untenable because it would require attributing everything from the origin of human beings to the daily rising of the Sun and the flight of an airplane to divine intervention; there would be no purpose in pursuing any aspect of science. This is not an acceptable option. Science works in so many instances that it cannot be rejected out of hand either piecemeal or in its totality.

However, Miller is also religious. His God did not create a universe that requires continual intervention in order to operate, much as a mechanic might have to continually tinker with an automobile to keep it running smoothly. Rather, his God created a universe whose components are governed by the natural laws of the universe. These laws operate to create galaxies and stars at one end of the size continuum and to create atoms, molecules, and living organisms at the other end. Miller argues that humans can rejoice in the existence, love, and beneficence of God only if they have the free will to either embrace or reject the existence of God. And they can have this free will only if there is no inarguable proof of God's existence. In a world in which God created every species by a separate act of creation, the inability of natural forces to produce life would eventually be discovered by curious humans and would prove the existence of God, robbing humans of the option to not believe. Thus, Miller sees his God in the processes of evolution and in the other forces that operate in nature.

Can we see God in astrobiology if we so choose? Astrobiology brings the complete range of physical and biological sciences together under a single coherent umbrella of mutually interactive ideas. Each discipline is important in its own right, yet each also relates to the others in complex and delightful ways. There is order to the universe as a whole, as well as to each part separately. Through astrobiology we can explore our own connections, individually and collectively, to this world. We do so locally in understanding terrestrial ecosystems, climate, and the history of life, and on a larger scale by understanding the ways in which biota interact with their planetary environment, the range of different environments in which they can survive, and the multiplicity of biochemical processes that allow them to thrive in these varied environments. And we do so at very much larger scales by understanding the objects in our solar system—their formation, evolution, history, and potential for life—and by understanding other stellar and planetary systems and what they might contain. We can see the world as a coherent system of physical and biological processes and entities, each relating to all of the others.

For those who are so inclined, it is easy to find God in the world around us. At the same time, the search for natural explanations in cause and effect and in understanding the origin and evolution of life also allows the absence of a God. As Miller says, "What science cannot do is to assign either meaning or purpose to the world it explores." It is up to us, individually and collectively, to find the meaning in the world and to choose our individual belief systems.

7

The Two Cultures of Astrobiology

ASTROBIOLOGY AS A SCIENTIFIC DISCIPLINE TOUCHES IN significant ways on the larger society. Scientists working in this field do research that touches on issues with strong intellectual and emotional connections to the public. I have explored some of these connections here, addressing the questions of how we do science, why we explore, how federal science policy supports our efforts, and the relationship between science and religion. Despite the importance of these connections, scientists often make little effort to discuss their work as it relates to society, either within their own community or in serious discussions with the public. Some scientists even question the value of doing science that is of deep interest to the public, unless it also addresses the questions that are of most interest to the individual scientists themselves or to the science community.

Should our exploration of the solar system include determining what makes one planet habitable and another uninhabitable? Should the search for life in our solar system, possibly on Mars or Europa, be an important part of our planetary research program? More broadly, what are the connections between astrobiology and the public, and between astrobiologists and the public? How do the scientists view the discipline of astrobiology, and how does the public view it? How should the science community interact with the public, and what are the appropriate forums for doing so? Finally, once we do recognize the value in understanding the relationship between science and society, what should we do about it? What changes does the scientific community need to make to be able to respond to some of these issues? While every interaction involves at least two individuals or groups, what can and should the scientists do to encourage and enhance these interactions?

Is the Search for Life a Valid Research Focus for NASA?

The public has demonstrated time and again its interest in knowing whether there is life elsewhere. The tremendous appeal of Carl Sagan's *Cosmos* television series, shown originally two decades ago and still replayed today, shows this interest, as does the excitement generated by discoveries that relate to life on other planets. The success of blockbuster movies like *Independence Day*, *Men in Black*, and the *Star Wars* and *Star Trek* series also shows the public's desire to think about issues related to extraterrestrial life.

The science community, on the other hand, has been somewhat ambivalent about issues related to life elsewhere. A significant number of scientists reject extraterrestrial life as a valid or suitable research topic. Given the variety of opinions, what priority should NASA assign to extraterrestrial life issues in setting its research agenda?

Before trying to reach any conclusions in this regard, let us consider some of the issues and concerns that have been raised by some of the scientists. The objections have come perhaps most explicitly from within the planetary science community, in that a search for life (on Mars or Europa) is largely a component of that discipline. A number of scientists in that field believe that the space exploration program should encompass much more than a search for life elsewhere, and that while astrobiology could be a small part of the overall program, it should not be a central theme.

Astrobiology is much more than just the search for living organisms, however. The major goal of astrobiology is to understand the origin (and possible origins), evolution, and distribution of life in the universe. In this context, we must understand what characteristics make a planet habitable and how the geological, geophysical, and geochemical processes that take place on planets lead to a planet being either habitable or uninhabitable. That is, we have to understand the processes that produced the architecture both of our solar system and of other solar systems. This approach encompasses much of planetary science, in that the formation and evolution of each planet, and of the entire suite of comets, asteroids, and Kuiper-belt objects, must be understood. Understanding the underlying processes will allow us to know the significance of a discovery of life on Mars or Europa, or the potential for life to exist outside our solar system. The search for life on Mars or Europa is important for understanding habitability, especially in that the presence or absence of life helps us to define habitability, but it is not an isolated goal in itself.

A related concern is that, if we should explore Mars or Europa and not find life, an exploration program based on such a search might cease to be of interest to the public and might consequently be scaled back dramatically. The history of Mars exploration is used as an example. The failure of the Viking spacecraft in the mid-1970s to find life seemed to bring Mars exploration to a halt for two decades. From the scientific perspective, certainly, a finding of no life on Mars or Europa would be just as important as a finding of life, because either would add to our understanding of what makes a planet habitable and what the actual distribution of life in our solar system is. We need to ensure that the Mars program is sold not as "finding life on Mars" but instead as "finding out if habitable conditions ever existed and *whether* there is life on Mars," and it must be clear that the implications of either result are profound.

The reference to the Viking missions may not be as valid as is commonly thought. A major factor in the twenty-year hiatus in Mars exploration following Viking was the absence at that time of any well-articulated scientific rationale for continuing to explore Mars and of any mission concepts that would address important scientific questions. The exploration *did* pick up again after the scientific rationale and mission plans were constructed.

In addition, the tremendous increase in the pace of Mars exploration in the 2000 to 2004 time frame not only coincides with the reappraisal of Mars as a potential abode for life but is a direct result of that interest. The funding increases directly followed the interest and excitement about potential life. The current Mars program emphasizes the search for life as a central theme, but it is just one of several themes. The implementation of the program addresses key issues in Mars's geological and geophysical history, climate, and atmosphere in addition to looking for life. This breadth is important both for understanding Mars as a planet or system and for determining the planetary context of the astrobiological investigations.

Finally, some scientists object that there is little scientific justification for thinking that life could be present on Mars or elsewhere and claim that the astrobiology community's use of this argument to justify an expanded exploration program is at best exaggerated and misleading to the public (and at worst disingenuous). In contrast, though, is the view that astrobiology is an intellectually stimulating endeavor of enormous scientific value, with direct relevance to how scientists address questions in solar system exploration. The intellectual arguments behind suggestions that life might exist (or might have existed in the past) on other planets have emerged from an increased understanding of both terrestrial biology and planetary environments. The very early origin of life on Earth, the ability of life to thrive in a very wide range of environments, the possibility that life originated in "extreme" environments, and the fact that these environments are likely to exist on other planets all seem to indicate that life could have had an independent origin elsewhere or could have been transferred elsewhere from Earth.

In addition, asking about the distribution of life in the universe allows us to unite disciplines within space exploration that otherwise are only weakly connected. We can ask whether there are Earth-like planets around other stars and what the implications of the characteristics of the Jupiter-like planets that have been discovered are for the existence of habitable planets. These questions provide an integrating theme between understanding our own solar system and understanding the origin of other planetary systems and their connections to star formation. Our own solar system becomes the one example that we know in the most detail, and the other systems become our means to understanding planetary systems in general. The composition of the interstellar medium, the formation and history of the galaxy, and even the earliest history of our universe all have implications for star and planet formation and composition, and thereby for planetary habitability. Many of the activities that are taking place within space science have implications for astrobiology, and astrobiology can serve as the integrating theme that ties them all together. The discipline of astrobiology is about exploring these logical inferences and trying to find out whether we have a full picture of the connections between the universe, galaxies, stars, planets and, in the end, life.

Determining whether humans are alone in the universe certainly is of profound importance in terms of how we view ourselves as a species, as a society, and even as a biosphere. Revolutions in the 1990s in terrestrial biology, paleontology, astrophysics, and planetary science have brought us to the verge of a possible discovery of extraterrestrial life on Mars or Europa, if it exists there. We don't know what we will find when we really explore the most likely potential habitats on these planets, but it is now within our capability to explore them. Even if we find no life, we will have learned a tremendous amount about how these planets and satellites formed and evolved, about their similarities to Earth, and about the processes that determined the architecture of our solar system. Understanding this architecture will tell us about the broader context in which the Earth resides and about our own existence here. It also will allow us to extrapolate to understanding the distribution of Earth-like planets that might exist in other planetary systems, orbiting other stars.

Questions such as these inspire the public's interest and imagination. One of the hallmarks of Western civilization is this desire to understand the world. Certainly, we have been an "exploring" society for most of the last two thousand years, and looking for our place in the universe is one of the activities that give meaning to life. Exploring the Earth and understanding the origin and history of life, exploring the solar system and determining whether there is other life in our own backyard, and exploring the universe in order to understand our relationship to it are all key parts of being an exploring society.

Today, we look back five hundred years and see that Europeans' discovery of the New World and the subsequent global exploration and expansion that took place were watershed events in human history. Five hundred years from now, people will look back on this time as the time when humans first left the surface of the Earth and ventured into space, first visited the Moon, and first explored Mars and beyond. Addressing questions that have been at the center of intellectual thought since the beginning of recorded history has to count as one of the most valued and noble activities in which we could engage.

This convergence of scientific and intellectual thought is ultimately the strongest justification for continuing our space exploration program. The overlap between public excitement, interest, and enthusiasm and scientific interest in profound problems makes astrobiology an exciting way for the public and the scientists to come together. And it is the justification for astrobiology's importance in space exploration.

C. P. Snow's "The Two Cultures"

The ability of astrobiology to bridge the scientific enterprise and the philosophical and intellectual world, while not unique, places it among an important group of scientific disciplines. We can use it as an example of how science fits into our society and how our society views science, and we can use it as a way to engage the public in learning about the nature of science. Nonscientists tend to see science as a monolithic enterprise that is distinct from all other endeavors in which we as a society engage. Similarly, scientists also tend to see their work as distinct from the rest of society. These perspectives come out, for example, in our university educational system, where those who are trained in the humanities receive a minimal science education, and those who are trained in the sciences receive a minimal humanities education. In the end, the lack of attention scientists give to social and cultural issues stems both from this minimal education and from a federal science policy that (as discussed earlier) places relatively little value on making these connections.

The issue of the relationship between science and other intellectual endeavors bubbled up into public and intellectual consciousness during the height of the Cold War, when the federal policy for science was still relatively new. In 1959, C. P. Snow gave a public lecture (the Rede lecture) at Cambridge University in England that was published in pamphlet form the following day. The title of his talk was "The Two Cultures," and the gist of it was that intellectuals in modern society can be divided into two camps—those who are scientifically literate but culturally ignorant and those who are culturally literate but scientifically ignorant; that these groups are essentially unable to interact effectively with each other; and that we are all the poorer for this state of affairs. Snow was in a relatively rare position to comment on the distinctions between these two groups. He had been trained originally as a scientist and had spent time working in science laboratories, had made a transition to writing novels and had circulated amongst the literary elite of his era, and had also held a senior civil service position in the British government.

In a book published in 1993, Snow elaborated on the issues he raised in his lecture:

In our society (that is, advanced western society) we have lost even the pretence of a common culture. Persons educated with the greatest intensity we know can no longer communicate with each other on the plane of their major intellectual concern. This is serious for our creative, intellectual and, above all, our normal life. It is leading us to interpret the past wrongly, to misjudge the present, and to deny our hopes of the future. It is making it difficult or impossible for us to take good action.

I gave the most pointed example of this lack of communication in the shape of two groups of people, representing what I have christened "the two cultures." One of these contained the scientists, whose weight, achievement and influence did not need stressing. The other contained literary intellectuals. I did not mean that literary intellectuals act as the main decision-makers of the western world. I meant that literary intellectuals represent, vocalize, and to some extent shape and predict the mood of the non-scientific culture: They do not make the decisions, but their words seep into the minds of those who do. Between these two cultures—the scientists and the literary intellectuals—there is little communication and, instead of fellow-feeling, something of hostility.

He added: "It is dangerous to have two cultures which can't or don't communicate. In a time when science is determining much of our destiny, that is, whether we live or die, it is dangerous in the most practical terms. Scientists can give bad advice and decision-makers can't know whether it is good or bad."

Snow's lecture had come at a particularly poignant time. Nuclear weapons had only recently been created and used, the Cold War was nearing its peak, and there was a constant threat of it turning into a deadly Hot War. The Soviets had launched the first Sputnik into orbit less than two years earlier, giving the entire globe a legitimate fear about nuclear weapons in space. Snow recognized the contributions that an enhanced attention to science and technology could make in solving some of society's problems. In particular, he recognized a global need for more and better-trained scientists, especially those who could grasp and respond to social concerns.

Snow's emphasis at the time was on educating the literary intellectuals on the nature of science and technology, and today his work may seem somewhat condescending toward nonscientists. Of course, the zeitgeist of the time was such that this was the immediate concern. The Soviets' launch of Sputnik had created a panic in the United States over the potential technology gap between the East and the West, along with an urgent call for more and better education in science, math, and engineering.

The same idea of distinct cultures holds today as well. In college, which is the last time that students get a formal education that might cross over the discipline boundaries, students are for the most part poorly educated outside their major discipline. Students in the humanities can actually graduate without having a good concept of what science is, how it works, and what its relationship to social concerns is or should be. And students in the sciences have little opportunity to discuss the role of science in society, why we do science, or what the social responsibility of scientists is.

Today, we can read Snow's book and recognize the necessity for both sides to understand the broader context of their work. We can understand the need to take an integrated approach that crosses the boundaries between disciplines and, in the particular context of our discussions here, to examine and value the connections between science and the broader issues and the role that science plays in our society. This need applies both to scientists and to nonscientists. Certainly, nonscientists need an understanding of science, if only to be able to make well-informed decisions in an increasingly technological society. And scientists need to know how their science affects society, so that they are not doing their science in isolation. Only if each of these two cultures understands something of the other can science be an integral component of our society.

This lack of crossover is particularly interesting in the light of a couple of issues that I have already noted. One involves the dichotomy between science as a way of addressing specific needs or solving specific problems in our society—or, in the case of basic science, preparing the groundwork that will lead to addressing them—and science as exploration and as a hallmark of civilization.

A second issue involves our underlying theme of using astrobiology to bridge the divide between Snow's two cultures. In fact, we have considered the value of crossing it in both directions. We wish to use astrobiology as a way of educating the nonscientists about the nature of science as an intellectual enterprise, how it operates, and what its uncertainties are. We also can use astrobiology to talk to the scientists about a role of science in society that goes beyond creating new products and solving specific problems, and as a way of focusing on the broader issues in society.

Scientists today are not trained to address the connections between their research and broader societal issues. Only a few universities offer courses in science and society, and these are generally not a mandatory part of the science curriculum. Science majors are not asked to take courses in philosophy of science or to bring their science background into their humanities courses (or, often, even to take humanities courses). And they are not asked to read such seminal works as Vannevar Bush's *Science: The Endless Frontier*, Snow's *The Two Cultures*, or Pope John Paul II's *Faith and Reason*.

A second reason is equally important, and not unrelated to the first. During the post–World War II era of support for science, scientists did not emphasize or even mention an explicit understanding of the relationship between science and society. Researchers were trained with the viewpoint that the practice of good science and technology development were valid and valued ends in themselves; there was no need to integrate the results with the rest of society. Today's researchers were trained by those who lived their entire professional lives with that attitude, and the lack of discussion of these issues results in their promulgating the same attitudes.

We see this attitude today in the reluctance of many scientists to interact with the public in general or to discuss the societal implications of their science. When there is interaction it is often one-directional (that is, scientists talking *at* the public) rather than taking the form of a dialogue or discussion.

As an example, I attended a large international conference several years ago and had the opportunity to chat with a few colleagues about my interest in the connections between science and society. My colleagues were interested enough to listen to me discuss these ideas during a coffee break. One of them, a senior scientist in the field, asked if I was worried that talking about these issues would cause my colleagues to think I was "going soft on science." He might have been telling me either that this was his opinion or that it was likely to be the opinion of others. Or he might have been telling me that spending time on societal issues would take me away from doing science and would "marginalize" me in the eyes of my fellow scientists. Either way, it was an expression of the idea that addressing societal issues is not an important way for a scientist to spend his time.

The conclusion that I draw from this and other examples is that the science community as a whole has relatively little interest in carrying on a dialogue about these issues or in crossing over the boundary between the two cultures. At the least, the opportunity to spend time addressing these issues pales in comparison to spending time discussing the science itself.

This view can be seen in attitudes toward public outreach as well. Outreach involves taking the results of our scientific research back to the public rather than carrying on a discussion or dialogue with them. It is the first step in having significant interactions with nonscientists. There is a general sense that outreach is a valued and important activity for scientists to be involved in, at least as long as one spends only a small fraction of one's time on it.

The best example of a scientist who also engaged in outreach is Carl Sagan. Until his death in 1996 he was America's most visible public spokesman for science. His *Cosmos* television series and book explored science and its connections to the public interest and emphasized a truly interdisciplinary and integrated approach to science and society. Sagan wrote a number of books aimed specifically at the public—and was pilloried for it by many scientists. They recognized the value of his research results (which had a substantial impact on the science community that continues today), but many also denigrated his emphasis on outreach as pandering to the public interest. It is widely believed that this emphasis on communicating with the public and discussing social issues kept him from being elected to the National Academy of Sciences.

I can cite other, more recent examples on a smaller scale. One of my colleagues is actively engaged in outreach and spends a substantial fraction of her time interacting with the public (and the rest in doing science). She sees this as a valuable activity and is strongly committed to doing it. She is also reluctant for other scientists to know about it, afraid that people will see her as being less than a committed scientist, and that this will impede her career. It is of special concern because she is worried that a woman spending a substantial fraction of her time on outreach will be seen, especially, as "soft on science."

Another colleague, this one a senior person holding a tenured position at a major research university, decided to make a strong commitment to education and outreach. Today, he devotes a significant part of his time to these activities, and it is relatively widely known that he does so. Apparently, though, he gets a lot of comments along the lines of, "What happened? All your funding dry up?" Many of our colleagues have the attitude that outreach is something to spend time on only if it is not possible to spend the time doing science.

These attitudes exist in part because of the sheer inertia in overcoming and changing them. Those brought up in an era in which outreach was not an important part of science and in which anything that detracted from doing science was problematic to one's career find it hard to make the mental change and come to value outreach. In all fairness, I should add that many scientists who have done outreach as a significant activity for a couple of decades are finding this work increasingly valued. In fact, some have told me that they have never personally seen or felt the negative view described above. Clearly, there is a wide range of opinions and activities in this area.

Now, at least, outreach is beginning to be valued within the science community. NASA now requires that each major proposal have a component dedicated to outreach. As the evaluation of these outreach components is beginning to play a larger role in the overall proposal evaluations, scientists are starting to take them seriously. While the first rounds of outreach activities consisted of little more than creating Web sites, scientists now are partnering with experts in education and outreach, and collaborating with schools, school districts, and museums. Research organizations in space science are beginning to hire specialists in education and outreach who can help to coordinate activities among individual scientists, advise them on what ideas to put into their proposals and how to implement them, and create outreach programs tailored to individual research groups.

As we begin to value outreach, so may we soon begin to value the broader issues in society as a part of our science. There appears to be an interest on the part of some scientists, expressed privately, at least, in having some discussion of these issues. This is not too surprising for the younger scientists, in that many of them entered planetary science and astrobiology specifically because of the broader perspective of science discussed in Sagan's *Cosmos*. Today, Sagan is viewed somewhat as a visionary who paid attention to the public at a time when it was not something that others thought important.

The issues that Snow addressed in "The Two Cultures" have not disappeared in the years since 1959. At best, there is only a weak incentive to change. Certainly there is little pressure to change how we teach science, either to nonscientists or to scientists. And as long as tomorrow's teachers are taught by people who hold the same attitude—that science is separate from society—there will be no real change in how science is taught.

A Call for Action

The beginning of the twenty-first century has seen a fundamental "sea change": the public's interest in space exploration has converged with scientists' interest in the potential and actual distribution of life in the solar system and beyond. In an ideal world, that convergence would be followed by a complete shift in the approach that scientists take to interacting with the public. In fact, scientists are just beginning to recognize the value of such interaction, which is taking the form of outreach and both formal and informal educational opportunities. Most scientists now consider outreach a valid and valuable activity, even if the entire community has not quite accepted it. There would be tremendous value in a sustained dialogue with the public over the meaning of the results we obtain as well, but this type of activity is neither valued nor seen as particularly useful today. How should the astrobiology and space science communities change to respond to these issues? What is their responsibility for interacting with the public that supports them, and what are the appropriate ways to do so? I offer the following suggestions.

We need to educate the public on the nature of science and how it differs from other ways of knowing about our world. It is our responsibility as scientists and educators to ensure that we address the fundamental issues in science today. These issues certainly include understanding details about how the world works (such as the cause of the seasons or the structure of DNA), as is typically taught in science classes. But they also include understanding the nature of science, how science differs from other endeavors in what it can tell us about the world, and what the limits of science are. We are shirking our duty if we do not address these issues with the public, and doubly so if we then complain that the public does not understand what science is and how it works. Emphasizing discussion of these issues with the public is an extremely difficult task, however, in that most scientists do not even recognize that there is a problem or that we need to change the emphasis of what we teach and how we teach it. It probably is a truism that real change will take place only with a complete turnover in educators, over one or two generations, so that those who recognize the different approaches and issues can fully integrate them into their teaching programs. This turnover cannot even begin, however, until we recognize the need to change.

We need to change scientists' mind-set that science should be supported solely because it is "good." Science and technology development have received munificent federal funding for the past half-century. This support has allowed the best scientists to develop the attitude that they do not need to justify their research in terms of the broader societal interest. However, the era of unquestioned support for science and technology may be over. Science today often is seen by Congress, for example, as just one more constituency competing for an ever-shrinking pot of available federal funds. Unless we recognize this shift and respond appropriately, we will see a continued erosion of funding for the programs we value. We need to recognize that programs such as astrobiology and space exploration are being driven to a large extent by the public interest and excitement in them, and that the public has a legitimate and appropriate interest in what scientists are doing. Only if we address this interest can we expect to see continued support for large research projects.

We need to enter into a dialogue with the public on the role and significance of astrobiology rather than devaluing the perspectives of nonscientists. Although many scientists are beginning to value outreach and interaction with the public, too many do it because "if we tell the public about what we're doing, they'll keep sending us money." In the long run, we will not be successful at interacting with the public if we take the attitude that we know it all and that interaction between scientists and the public is a unidirectional monologue (we lecture to them and they listen to us). We are addressing questions in astrobiology and space exploration because they are of interest to the public, and the public has a legitimate and real interest both in the results and in participating in the exploration. Thus, we must not only communicate with them, we also must enter into a dialogue with them regarding the overall directions for the program and the high-level questions to be addressed. One implication of having a dialogue as opposed to a monologue is that, all of a sudden, we are putting the science community and the public on similar levels in terms of influencing the program. The public become peers who "buy into" the program at a deep level, and the scientists have to come down off the pedestal they have occupied for the last fifty (or, arguably, three hundred) years. Science is a part of society, and scientists are a part of society, and we should run the program in a way that recognizes this reality.

We need to engage scientists in a discussion of the role of science in society and persuade them to value the "exploration" and "philosophical" aspects of planetary science and astrobiology. It will not be possible to interact effectively with the public on the role and value of space exploration and astrobiology research unless scientists, as a community, understand and appreciate these issues themselves. Thus, we need to begin to value the "nonscience" aspects of the issues, and to understand and develop them ourselves. Our university courses must include discussions of why we do science and what the societal context and implications of our science programs are. This need applies to undergraduate nonmajors courses in which we reach the future citizens of our country, but it also applies to upper-level undergraduate and graduate courses in which we teach our future professionals and educators. In order to do this effectively, we need to begin by addressing these questions within the science community. Only after we, as scientists, have discussed these questions will we be able to communicate their value and interest to our students and to the public. We need to include discussion of these issues, for example, within our science conferences and as a central part of our research activities. Just as most scientists understand that the research is not complete until the paper is written and published, we also have to understand that it is not complete until we have explained it to the public and engaged them in a dialogue on its broader values and significance.

We need to recognize the importance of outreach in the broadest sense and to value the contributions of those who engage in it. We all know scientists who are not able to interact with the public, who cannot communicate effectively with those outside their own very narrow environment, and who do not appreciate the value of interacting with the public. Not everybody should be required to participate in outreach, but we need to value the contributions of those who do interact with the public and not marginalize them for their activities. To do so, we need to fund effective outreach in its broadest sense. Outreach and education mean more than the standard "K through 12" teacher-training workshops. Scientists who write books and articles aimed at the public, who give public talks either locally to Rotary Clubs and other civic groups or in broader venues, or who interact with the media and thereby reach large numbers of people all need to be supported under the heading of outreach and education. Organizations that employ scientists, such as government laboratories, NASA field centers, and universities, need to recognize and value contributions in these areas.

We have to do these things while continuing to do the highest-quality science. We need to work with the public to define the overall goals and directions of the space science program. At the same time, the science community is the group best able to determine what the best science is, the appropriate strategy to address the science goals, how best to implement the strategy, and when the technology is at an appropriate level to allow us to fly capable missions that will do high-quality science. The National Research Council constructs strategies to explore the Earth, the solar system, and beyond. NASA is very good at using these strategies and developing implementation plans that run exciting missions that do high-quality science. We should take advantage of these existing programs, but we should also involve the public in setting the high-level goals and objectives. Doing so will complete the circle and allow us to do high-quality science that has real meaning to the public and to society as a whole. We want to involve the public as an equal partner, but that does not mean that we have to fly missions to explore the face on Mars! As scientists, it is our responsibility to ensure that we are carrying out a program of high-quality and high-priority science.

Final Comments

We are exploring our solar system and trying to understand the origins and distribution of life here and beyond in order to satisfy a deeply embedded human curiosity about the world. In the end, what does our exploration really mean? The following quote about the potential for life elsewhere (variously attributed to Bertrand Russell, Carl Sagan, and *Pogo* cartoonist Walt Kelly but actually traceable to science fiction author Arthur C. Clarke) sums it up nicely: "There are two possibilities. Maybe we're alone. Maybe we're not. Both are equally frightening."

It would be truly intimidating to determine that we are the only life —or the only intelligent life—in our galaxy. That knowledge would have truly profound implications for understanding our connection to the world around us and for understanding the nature of humanity and what it means to be human. At the same time, it would be just as intimidating to find even a single example of non-Earth life on any other planet, whether it be the simplest conceivable microbe or intelligence with whom we might (or might not) be able to communicate. The knowledge of life's existence elsewhere, by itself, will change how we view ourselves, our planet, and the universe as a whole.

T. S. Eliot summed up the implications and rationale for astrobiology very effectively when he wrote in *Little Gidding*: "We shall not cease from exploration, and the end of all our exploring will be to arrive where we started and know the place for the first time." Understanding the universe and the potential and actual distribution of life within it will illuminate our own existence here on Earth and help us to comprehend our own species, our own society, and our own individual lives.

References and Additional Reading

Barbour, I. G. *When Science Meets Religion.* New York: HarperCollins, 2000.

Benner, S. A., K. G. Devine, L. N. Matveeva, and D. H. Powell. "The missing organic molecules on Mars." *Proceedings of the National Academy of Sciences* 97 (2000): 2425–2430.

Bennett, J., S. Shostak, and B. Jakosky. *Life in the Universe.* San Francisco: Addison-Wesley, 2003.

Blackburn, S. *Think: A Compelling Introduction to Philosophy.* Oxford: Oxford University Press, 1999.

Boss, A. P. "Extrasolar planets." *Physics Today* 49 (1996): 32–38.

Bracewell, R. N. *The Galactic Club: Intelligent Life in Outer Space.* San Francisco: W. H. Freeman, 1974.

Brasier, M. D., et al. "Questioning the evidence for Earth's oldest fossils." *Nature* 416 (2002): 76–81.

Brock, T. D., M. T. Madigan, J. M. Martinko, and J. Parker. *Biology of Microorganisms.* 7th ed. Englewood Cliffs, N.J.: Prentice-Hall, 1994.

Bush, G. W. "A Renewed Spirit of Discovery: The President's Vision for U.S. Space Exploration." The White House, 2004.

Bush, V. *Science: The Endless Frontier.* Washington, D.C.: Government Printing Office, 1945.

Butler, R. P., and G. W. Marcy. "A planet orbiting 47 Ursae Majoris." *Astrophysics Journal* 464 (1996): L153–L156.

Calvin, W. H. "The emergence of intelligence." *Scientific American* 271 (1994): 100–107.

Campbell, N. A. *Biology.* 3rd ed. Redwood City, Calif.: Benjamin/Cummings, 1993.

Carr, M. H. *The Surface of Mars.* New Haven: Yale University Press, 1981.

———. *Water on Mars.* New York: Oxford University Press, 1996.

Carr, M. H., and J. Garvin. "Mars exploration." *Nature* 412 (2001): 250–253.

Chalmers, A. F. *What Is This Thing Called Science?* 3rd ed. Cambridge: Hackett, 1999.

Chang, S. "Planetary environments and the conditions of life." *Philosophical Transactions of the Royal Society, London,* A 325 (1988): 601–610.

Cleland, C. E. "Historical science, experimental science, and the scientific method." *Geology* 29 (2001): 987–990.

———. "Methodological and epistemic differences between historical and experimental science." *Philosophy of Science* 69 (2002): 474–496.

Cocconi, G., and P. Morrison. "Searching for interstellar communications." *Nature* 184 (1959): 844–846.

Darwin, C. *On the Origin of Species*. A Facsimile of the First Edition. Cambridge: Harvard University Press, 1964.

Davies, P. *Are We Alone? Philosophical Implications of the Discovery of Extraterrestrial Life*. New York: Basic Books, 1995.

Dawkins, R. *The Blind Watchmaker*. New York: W. W. Norton, 1996.

Deamer, D. W., and G. R. Fleischaker. *Origins of Life: The Central Concepts*. Boston: Jones and Bartlett, 1994.

Des Marais, D. J., L. J. Allamandola, S. A. Benner, and 18 other authors. "The NASA Astrobiology Roadmap." *Astrobiology* 3 (2003): 219–235.

Dick, S. J. *The Biological Universe*. Cambridge: Cambridge University Press, 1996.

———. "Consequences of success in SETI: Lessons from the history of science." In *Progress in the Search for Extraterrestrial Life*, ed. G. S. Shostak, 521–532. San Francisco: Astronomical Society of the Pacific, 1995.

Ehlers, V. *Unlocking Our Future: Toward a New National Science Policy*. Committee on Science, U.S. House of Representatives. Washington, D.C.: Government Printing Office, 1998.

Feyerabend, P. "How to defend society against science." In *Introductory Readings in the Philosophy of Science*, 3rd ed., ed. E. D. Klemke, R. Hollinger, and D. W. Rudge, 54–65. Amherst, N.Y.: Prometheus Books, 1998.

Frankl, V. E. *Man's Search for Meaning*. New York: Simon and Schuster, 1959.

Gould, S. J. *Rocks of Ages: Science and Religion in the Fullness of Life*. New York: Ballantine Books, 1999.

———. *Wonderful Life*. New York: W. W. Norton, 1989.

Haynes, R. H., and C. P. McKay. "The implantation of life on Mars: Feasibility and motivation." *Advances in Space Research* 12 (1992): 133–140.

Hempel, C. G. "Studies in the logic of explanation." In *Introductory Readings in the Philosophy of Science*, 3rd ed., ed. E. D. Klemke, R. Hollinger, and D. W. Rudge, 206–224. Amherst, N.Y.: Prometheus Books, 1998.

Jakosky, B. *The Search for Life on Other Planets*. Cambridge: Cambridge University Press, 1998.

Jakosky, B. M., and M. P. Golombek. "Planetary science, astrobiology, and the role of science and exploration in society." *Eos, Transactions of the American Geophysical Union* 81 (2000): 58.

Jakosky, B. M., and R. J. Phillips. "Mars' volatile and climate history." *Nature* 412 (2001): 237–244.

Jakosky, B. M., F. Westall, and A. Brack. "Mars." In *Astrobiology*, ed. J. Baross and W. Sullivan. In press.

John Paul II. *Fides et ratio. On the Relationship between Faith and Reason*. Encyclical Letter to the Bishops of the Catholic Church. www.catholic-pages.com, 1998.

Joyce, G. Foreword to *Origins of Life: The Central Concepts*, by D. W. Deamer and G. R. Fleischaker, xi. Boston: Jones and Bartlett, 1994.

Kieffer, H. H., B. M. Jakosky, C. W. Snyder, and M. S. Matthews, eds. *Mars*. Tucson: University of Arizona Press, 1992.

Klein, H. P. "Did Viking discover life on Mars?" *Origins of Life and Evolution of the Biosphere* 29 (1999): 625–631.

Klein, H. P., N. H. Horowitz, and K. Biemann. "The search for extant life on Mars." In *Mars*, ed. H. H. Kieffer, B. M. Jakosky, C. W. Snyder, and M. S. Matthews, 1221–1233. Tucson: University of Arizona Press, 1992.

Knoll, A. K., and M. J. Osborne, eds. *Size Limits of Very Small Organisms: Proceedings of a Workshop*. Washington, D.C.: National Academy Press, 1999.

Kraft, C. *Flight: My Life in Mission Control*. New York: Penguin Putnam, 2001.

Kuhn, T. S. *The Copernican Revolution*. Cambridge: Harvard University Press, 1957.

———. *The Structure of Scientific Revolutions*. 3rd ed. Chicago: University of Chicago Press, 1996.

Lafleur, L. J. "Astrobiology." *Astronomical Society of the Pacific Leaflet* 143 (1941): 333–340.

Lederberg, J. "Exobiology: Approaches to life beyond the Earth." *Science* 132 (1960): 393–400.

Lissauer, J. J. "Planet formation." *Annual Reviews of Astronomy and Astrophysics* 31 (1993): 129–174.

Marcy, G. W., and R. P. Butler. "A planetary companion to 70 Virginis." *Astrophysical Journal* 464 (1996): L147–L151.

Mayr, E. *This Is Biology: The Science of the Living World*. Cambridge: Harvard University Press, 1997.

McMullin, E. "Life and intelligence far from the Earth: Formulating theological issues." In *Many Worlds*, ed. S. Dick, 151–175. Philadelphia: Templeton Foundation Press, 2000.

McGrath, A. E. *Science and Religion: An Introduction*. Malden, Mass.: Blackwell, 1999.

McKay, C. P. "Does Mars have rights? An approach to the environmental ethics of planetary engineering." In *Moral Expertise*, ed. D. MacNiven, 184–197. New York: Routledge, 1990.

McKay, D. S., E. K. Gibson Jr., K. L. Thomas-Keprta, and 6 other authors. "Search for past life on Mars: Possible relic biogenic activity in Martian meteorite ALH84001." *Science* 273 (1996): 924–930.

McSween, H. Y. Jr. "What we have learned about Mars from SNC meteorites." *Meteoritics* 29 (1994): 757–779.

Miller, K. R. *Finding Darwin's God: A Scientist's Search for Common Ground between God and Evolution*. New York: HarperCollins, 1999.

Miller, S. L. "The prebiotic synthesis of organic compounds as a step toward the origin of life." In *Major Events in the History of Life*, ed. J. W. Schopf, 1–28. Boston: Jones and Bartlett, 1992.

Morris, S. C. *The Crucible of Creation*. New York: Oxford University Press, 1998.

National Aeronautics and Space Administration. *2003 Strategic Plan*. Washington, D.C.: NASA, 2003.

National Aeronautics and Space Administration. Exobiology Program Office. "An exobiological strategy for Mars exploration." *NASA SP-530*, 1995.

National Research Council. *New Frontiers in the Solar System: An Integrated Exploration Strategy*. Washington, D.C.: National Academy Press, 2003.

Nealson, K. H. "The limits of life on Earth and searching for life on Mars." *Journal of Geophysical Research* 102 (1997): 23675–23686.

Polanyi, M. "The republic of science: Its political and economic theory." In *Criteria for Scientific Development: Public Policy and National Goals*, 1–20. Cambridge: MIT Press, 1968.

Popper, K. "Science: Conjectures and refutations." In *Introductory Readings in the Philosophy of Science*, 3rd ed., ed. E. D. Klemke, R. Hollinger, and D. W. Rudge, 38–47. Amherst, N.Y.: Prometheus Books, 1998.

Race, M., G. Schwehm, S. Shostak, and 4 other authors. "Societal, legal, educational and communication issues in the search for extraterrestrial life." *Advances in Space Research*, in press.

Ruse, M. *Can a Darwinian Be a Christian?* Cambridge: Cambridge University Press, 2001.

Ruse, M., ed. *Philosophy of Biology*. Amherst, N.Y.: Prometheus Books, 1998.

Sagan, C. *Cosmos*. New York: Random House, 1980.

Schopf, J. W. "Apex Chert and the antiquity of life." *Science* 260 (1993): 640–646.

Schopf, J. W., A. B. Kudryavtsev, D. G. Agresti, and 2 other authors. "Laser-Raman imagery of Earth's earliest fossils." *Nature* 416 (2002): 73–76.

Schrodinger, E. *What Is Life?* Cambridge: Cambridge University Press, 1944.

Segura, T. L., O. B. Toon, A. Colaprete, and K. Zahnle. "Environmental effects of large impacts on Mars." *Science* 298 (2002): 1977–1980.

Shils, E., ed. *Criteria for Scientific Development: Public Policy and National Goals*. Cambridge: MIT Press, 1968.

Shock, E. L. "High-temperature life without photosynthesis as a model for Mars." *Journal of Geophysical Research* 102 (1997): 23687–23694.

Snow, C. P. *The Two Cultures*. Introduction by Stefan Collini. New York: Cambridge University Press, 1993.

Sullivan, W. *We Are Not Alone: The Continuing Search for Extraterrestrial Intelligence*. Rev. ed. New York: Penguin Books, 1994.

Toulmin, S. "The complexity of scientific choice II: Culture, overheads, or tertiary industry?" In *Criteria for Scientific Development: Public Policy and National Goals*, ed. E. Shils, 119–133. Cambridge: MIT Press, 1968.

Van Doren, C. *A History of Knowledge*. New York: Ballantine Books, 1991.

Weinberg, A. M. "Criteria for scientific choice." In *Criteria for Scientific Development: Public Policy and National Goals*, ed. E. Shils, 21–33. Cambridge: MIT Press, 1968.

———. "Criteria for scientific choice II: The two cultures." In *Criteria for Scientific Development: Public Policy and National Goals*, ed. E. Shils, 80–91. Cambridge: MIT Press, 1968.

Westall, F. "The nature of fossil bacteria: A guide to the search for extraterrestrial life." *Journal of Geophysical Research* 104 (1999): 16437–16451.

Wetherill, G. W. "Formation of the Earth." *Annual Reviews of Earth and Planetary Science* 18 (1990): 205–256.

Wetherill, G. W. "Provenance of the terrestrial planets." *Geochimica et Cosmochimica Acta* 58 (1994): 4513–4520.

Wicander, R., and J. S. Monroe. *Historical Geology: Evolution of the Earth and Life through Time.* 2nd ed. San Francisco: West Publishing Company, 1993.

Wilson, E. O. *Consilience: The Unity of Knowledge.* New York: Random House, 1998.

Woese, C. R. "Bacterial evolution." *Microbiology Review* 51 (1987): 221–271.

Ziman, J. "What is science?" In *Introductory Readings in the Philosophy of Science*, 3rd ed., ed. E. D. Klemke, R. Hollinger, and D. W. Rudge, 48–53. Amherst, N.Y.: Prometheus Books, 1998.

Index

About the Author

BRUCE JAKOSKY is a professor and the Associate Director for Science in the Laboratory for Atmospheric and Space Physics at the University of Colorado at Boulder and a member of the Department of Geological Sciences. He has been at the university since 1982, when he received his Ph.D. in planetary science from the California Institute of Technology. He teaches undergraduate and graduate courses in terrestrial geology, planetary science, astrobiology, and extraterrestrial life. His research interests are in the geology of planetary surfaces, the evolution of the martian atmosphere and climate, the potential for life on Mars and elsewhere, and the philosophical and societal issues in astrobiology. He has been involved with the Viking, *Solar Mesosphere Explorer*, *Clementine*, *Mars Observer*, *Mars Global Surveyor*, and *Mars Odyssey* missions, and will be involved with the upcoming Mars Science Laboratory rover. He heads up the University of Colorado's team in the NASA Astrobiology Institute. He is the author of *The Search for Life on Other Planets*, a coauthor of a recent undergraduate textbook on astrobiology titled *Life in the Universe*, and a coeditor of *Mars*. He has published more than one hundred papers in the scientific literature, most of them dealing with various aspects of Mars. He lives with his wife, stepchildren, dog, and cat in Boulder, Colorado, where he has recently taken up skiing.